BEYOND EARTH

MAPPING THE UNIVERSE

EARTH

Edited by
David DeVorkin

WASHINGTON, D.C.

CONTENTS

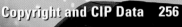

INTRODUCTION

David DeVorkin

A new permanent exhibition titled *Explore the Universe* opened quietly at the National Air and Space Museum on September 21, 2001, in the wake of the events of the 11th. More than ten years in planning, overlapping with six years in definition and development and hardly one year in full-scale production, it carries the simple but powerful message that as we have created new tools and techniques to explore the universe, our understanding of the universe has changed in dramatic ways. And our understanding is likely to continue changing as we acquire new tools of perception. This book of illustrated essays is meant to complement the exhibition as a personal accounting of the challenge and spirit of living in this universe, of recognizing that there is indeed a realm fit for study "Beyond Earth," and having the privilege to be able to explore it.

The exhibition itself concentrates on Western thought and human accomplishment. This book goes beyond the walls of the museum to explore how people of diverse backgrounds, cultures, and eras have discovered and described the cosmological state they have found themselves in, as they either study other cultures, practice science in their own culture, or go about their daily lives aware that they are part of a larger world, a world beyond the senses and beyond certainty. The authors have been asked to explore how their work, or the work of their subjects, has influenced their lives, their colleagues' views, and even the rest of the culture in which their science emerged. The result, I hope, is a suite of varied personal explorations of cosmology, some of which link scientific imagery to culture in the broadest sense and others that look at how culture has been somehow shaped by an awareness of the universe. Here we look at cosmology through the eyes of different cultures and at culture seen through the eyes of historians and cosmologists.

These essays are in large part personal visions by scholars who in their work have shown an interest in searching out visual metaphor in astronomical maps, world models, social structures, and computer data derived from a wide range of sensory devices. Underlying this effort is an interest in linking modes of visual representation to knowledge in science, but the thrust of the essays will be largely synthetic, speculative, and either broad ranging or anecdotal, but not heavily analytical nor encyclopedic. They will be, above all, personal statements, personal explorations.

The essays in this first section are historical, ethnographical, and anthropological, describing worldviews—or cosmologies—as they emerged in different cultures and at different times. Some comment on the interplay of cosmological worldview and the culture that created it. But they all are, moreover, products of personal and professional intellectual journeys—those taken by the writers as they explored questions they raised about the world around them. In this way we have provided a collective portrait of our own personal, intellectual pursuits and how we, as individual scholars, grew or changed as a result.

Authors have been chosen who have taken an intellectual journey and who can write engagingly about it in ways that, we hope, will reach a wide audience. These essays are narrative journeys taken by interesting people who are doing interesting things. They are not biographical profiles, but vignettes, stories about what it feels like to discover something new about the universe and about cultures that have pondered the universe in different ways. Some of the essays will let us take a peek at our craft, as scientist or historian, and how we acquired questions and sought out answers. Most of us have been fortunate enough to experience the thrill, either directly or indirectly, of encountering new universes. Our job here is to communicate the excitement and wonder of the encounter. This is payback time in a culture that permits and even applauds the efforts of those who choose intellectual journeys of discovery.

The book divides easily into three sections. The first grouping of essays explores our growing awareness of the universe as both a physical and a spiritual place. Our focus is on Western culture, and on how our sense of place has changed over time as we sought ways to predict the behaviors of things seen in the heavens. In reviewing our own history, we must be keenly aware of how our ever increasing ability to observe and to model the physical universe as a mechanism, or now as a set of rules, has actually removed us from the root questions that set us on this long journey of discovery. To remind us of these all-important roots to humanity, we have added several essays that explore both contemporary and historical cultures that retain a deep attachment to the spirit of the universe. If nothing else we hope this will remind all of us that the universe is not, as astrophysicist Sandra Faber likes to say, "out there. It is right here." We hold a portion of it in our very being.

This book was in large part stimulated by the presence of an exhibit and derives much of its flavor from it. Most of the authors were in one way or another involved as advisors to the exhibition team, and many elements of the exhibitry have found their way into these pages. Eric Long's wonderful photography, David Romanowski's lyric scripting, Beatrice Mowry's elegant designs all have informed and illuminated the book. Many of the graphics, including the dust jacket cover art, are works commissioned for the exhibition. The suite of illustrations collected by Joan Mathys for the exhibition provided the basis for the visual materials presented here.

Transforming the exhibition into a book has been managed with consummate skill and creativity by the editorial and photographic book staff of the National Geographic Society. Kevin Mulroy's enthusiasm for the project made it possible, Melissa Ryan's visual acuity made it live, Lyle Rosbotham made it fit together with delightful elegance, Marlis McCollum made it read right, and Lisa Thomas made it happen. It has been a privilege working with all of them.

SECTION I

THE
CLASSICAL
UNIVERSE

Planets and stars orbit the Earth in this 1660
model of the cosmos. Until the 17th century,
astronomers, philosophers, and laypeople
believed the Earth was the center of the universe.

THE
CLASSICAL
UNIVERSE

David DeVorkin

We start our journey with humans and human culture at the center of it all. Different cultures, of course, tended to place themselves at the center of the universe, but eventually they all realized that they occupied a single body called Earth. The essays in Section I describe variants on these most basic of universes. One of my hopes as acquiring editor was to show that there are strong affinities running throughout all cultures for regarding the sky in certain ways. Although the details vary widely, all cultures, at some time in their histories, tended to pound sticks into the ground or cliffs and watch as the shadows cast by the sun changed in length and position as the position of the sun changed in the sky. All cultures went beyond the daily cycle to find longer cycles that were built up into sophisticated systems of reckoning the passage of time. As the systems grew in sophistication, so did the instruments to observe them.

The discovery of time is prehistoric. In this section we present examples illustrating the richness of the many answers that mankind has devised to account for phenomena accompanying the passage of time, moving back and forth along the spectrum between sacred and profane. To fully appreciate these essays, we need to remember that preexisting spiritual forms (the sacred) were not necessarily replaced by contemporary forms (the profane) in all cultures. One can grow out of the other, as it has in Western culture. But this is not the way things must proceed. We can no more assume that to be the case than to reaffirm our world to be the center of the universe.

Sara Schechner has either managed or explored some of the world's greatest collections of antique scientific instruments. Here she reviews her encounter with the Earth-centered (geocentric) universes of classical antiquity, Islam, and the European Middle Ages. The Earth had been discovered; the problem was to make everything work nicely in attendance to its centrality. Schechner's introduction will serve as a springboard for appreciating not only the roots of cosmological inquiry, the personal and moral universes shared by all organized cultures—African, Asian, Native American, and European—but also the difficulties encountered by the latter culture as it acquired the means to observe the universe more closely, to the point where no geocentric model behaved nicely.

John S. Major will then transport us to China, where we will find, just as in all cultures, that control of the calendar was the hallmark of power and authority. Major has a deep interest in East Asian astronomy, astrology, and instrumentation, and has published extensively in the field. Major argues that China had astronomers before it had kings, which I take to mean that those with astronomical awareness, whether they were called shamans, priests, or astrologers, formed the nucleus of authority and power in the earliest organized families, bands, and tribes. What they did with this power defined a culture as the passage of time became formalized and codified.

Noted sky-lore enthusiast and ethnologist Von Del Chamberlain and art historian Christine Mullen Kreamer then take us to traditional societies, which, though widely removed on the Earth, share a commonality in the way they construct a moral universe that situates each person with respect to it. Theirs is less a physical space than a spiritual space, and each teaches us how power could be concentrated in life-forms and landforms familiar to people in specific climactic environments, and how the details of sky lore and calendar making may differ according to differences in these climes, all the while sharing common goals and a common association with the sky. One can well read through these essays and appreciate that the commonalities found between physical and cultural cosmologies must extend to all cultures.

To the Greeks, the sun, moon, planets, and stars were attached to transparent spheres surrounding Earth. The realm of the gods lay beyond the outermost sphere.

Deborah Jean Warner's reconnaissance of celestial mapmaking helps us appreciate how Western culture obtained a new level of order and predictive control on the universe. Maps became a way to describe the universe, requiring consensual decisions about the projective geometries to be employed, appropriate nomenclature, completeness, precision, standards of training, and use. Star maps were still very much a reflection of cultural roots in astrology and mysticism, but they also became scientific instruments, describing models of the universe that required skill to operate and interpret. Only as visual techniques became vastly augmented by the telescope did the requirements of the science surpass the limits of globes and maps. As we shall also learn in Section II, it was this increase in precision, continuing to use traditional methods, that finally led to the overthrow of the geocentric universe.

The last gasp of geocentrism in Western culture is the subject of Owen Gingerich's contribution, which ends Section I. Prominent author, historian, and astronomer, Gingerich has attained public notice as a deft interpreter of astronomical history. His quest to obtain a census of the remaining copies of the first two editions of Copernicus's great work has been a large and important mission in his life. Here he recounts why he took on the monumental task of documenting the diffusion of the book known as the key to the Copernican Revolution and where it has taken him.

ANCIENT COSMOLOGIES

Sara Schechner

In the fifth century B.C. the Greek philosopher Heraclitus observed about Nature's onward progression in time, "Everything flows and nothing abides; everything gives way and nothing stays fixed. You cannot step twice into the same river, for other waters go ever flowing on." But this is precisely what historians attempt to do. We try to return to that stream and feel the ancient waters and recognize how scientific theories and instruments did not just float down that stream but were immersed in it. The waters give life, and the reasoning, instruments, and world views of early scientists—indeed their cosmologies—should be understood within the context of the age and culture in which they lived. Remaining true to the very different worlds of three, four, or even five millennia ago is no small challenge, but is the only

Using a beaded string to measure the distances between the stars, a young astronomer maps the constellation Orion in this 19th-century woodcut. Careful observations of the sky over long periods of time with simple measuring tools provided the basis for cosmologies around the world.

way to interpret those past worlds satisfactorily for the present day, to make them live again.

In college I began my own personal search for that stream. I wanted to become a responsible physicist who was informed about the social, historical, and philosophical implications of my work. I understood that context was critical To this end, I began to take courses in the history and philosophy of science at Harvard in the 1970s and apprenticed myself at the Harvard Collection of Historical Scientific Instruments. Upon graduation, I decided to become an historian of science, a teacher, and a museum curator, for this path allowed me to combine my interests.

As with most historians, my forward path took me backward in time. What began as a personal search for order and purpose in modern physics became an exploration of earlier cosmologies. The natural philosophers of the past, I discovered, were no fools, nor was the public that had embraced their views.

ANCIENT EGYPT

In the Nile Valley and Delta more than 5,000 years ago, the ancient Egyptians personified the heavenly bodies as gods, and priests observed their risings and settings. The chief Egyptian god was Ra, the sun. Osiris was connected with the constellation we now call Orion. Sirius, the bright Dog Star that tags along behind Orion, was associated with Isis. Set was associated with the constellation we call the Big Dipper. The mythology of these gods was inextricably entwined with Egyptian religion, ritual, agricultural practices, and the development of practical astronomy.

A recurrent theme in Egyptian cosmology was the eternal cycle of birth, death, and rebirth. During the day, Ra, the sun god, was believed to cross the sky in a sailing vessel. Each night, he was swallowed by Nut, the sky goddess, whose starry body arched overhead. Each morning, Ra was reborn from her womb.

Birth, death, and rebirth also divided the Egyptian year into three seasons—the season of the annual Nile flood; the period of planting and growth; and the time of low water, when crops were harvested. During the season of the harvest, considered a sterile period, the constellation of Orion vanished from the sky, with the Dog Star, Sirius, not far behind. (In reality, Orion and Sirius were above the horizon during the day, but were hidden from view by the sun.) Seventy days later, Orion would return to the night sky, followed a few weeks later by Sirius, whose predawn rising coincided with the Nile's annual flood and the start of a new Egyptian year. Only the circumpolar stars—those circling the celestial North Pole—

never dipped below the horizon. Egyptians called these the "undying" stars.

The story of Osiris, Isis, and Set helped Egyptians interpret these astronomical and agricultural events. Osiris, the god of kingship and ruler of the dead, was also the god of the life-

Nut, the sky goddess, arches over the Earth god Geb in this ca 980 B.C. tomb painting of the Egyptian cosmos. Nut's hands define the western horizon; her feet, the east. Nut often appears in tomb architecture, defining the vault of the heavens.

ANCIENT COSMOLOGIES

sustaining Nile and a force behind the growth of vegetation. His sister, Isis, was the goddess of fertility, and Set was the god of chaos and sterility. In tombs, Osiris was often depicted as Orion sailing across the sky in a boat, with Isis as Sirius close behind him. Set was depicted as the circumpolar constellation Meskhetiu, or the Bull's Leg, which is our Big Dipper. According to legend, Set killed Osiris by trickery and floated his body down the Nile. Isis retrieved the body, hid it in the rushes, and magically conceived a baby by Osiris. Set discovered the hidden body, and in a fury tore it to pieces, which he scattered up and down the Nile. Isis recovered the dismembered body, mummified it for the proscribed period of 70 days, and so gave Osiris everlasting life. Just as Sirius had followed Orion from the sky, so Isis had followed Osiris down the Nile. During their absence, Set ruled because his constellation, the Bull's Leg, continued to circle the celestial pole. The land was sterile under Set's rule, but, just as Isis was able to produce a son from the dead Osiris and restore him to life, so too did the return of Sirius herald the revival of the land's fertility.

Observations of the stars, reinforced by religious belief and ritual, helped Egyptians predict the best times to plant and harvest. The need for stellar observations also stimulated the development of new methods of finding and measuring time. The earliest Egyptian calendar was lunar, but by the third millennium B.C. a 365-day solar calendar was adopted for civil reasons. Weeks were set at 10 days. By 2340 B.C. the Egyptians used the risings of key stars to divide the night into hours. These stars were selected because, like Sirius, they disappeared from the sky for 70 days out of the year. They are known as decans, because every 10 days a new dawn star marked the last hour of the night and the start of a new week. The sequence and spacing of decanal stars rising throughout the night led to the division of the night into 12 hours, a direct consequence of the decimal order of the Egyptian calendar. By 2000 B.C. this method was abandoned in favor of marking the stars' transit across the sky's meridian. By 1500 B.C. water clocks kept time when the stars were not visible. At about the same time, shadow clocks—now known as sundials—were also invented to divide the daylight into 12 hours, as had been done with the night.

Thus, Egyptian cosmology and civil order gave us our 365-day year and our 24-hour day. (The division of the hour into 60 minutes and the minute into 60 seconds was not Egyptian, but Babylonian, and of equally ancient origin.)

THE PHILOSOPHER'S UNIVERSE

Like the Egyptians, the early Greeks personified nature and attributed celestial and terrestrial events to the actions of a pantheon of deities. In Homer's *Iliad* and *Odyssey* and Hesiod's *Works and Days* the risings of constellations and bright stars announced the seasons for particular agricultural activities or forecast evil to mortals.

In the sixth century B.C., new philosophical modes of thought appeared alongside traditional Greek mythology. These philosophers asked new questions about the physical world: What was its shape? How was it made? How did things work? They required that the answers be found in nature, without recourse to gods.

The Greek philosophers believed that the cosmos was orderly and predictable. In spite of the appearance of diversity and change in the world, many believed that all things were made from a single type of matter. For Thales (ca 625–547 B.C. in Miletus), the primary matter was water; for Anaximenes (ca 546 B.C.), it was air; and for Leucippus (ca 440 B.C.) and Democritus of Abdera (ca 410 B.C.) matter was made from atoms. Other philosophers disagreed with these monistic notions. Some identified four elements: earth, air, fire, and water. Pythagoras (ca 560–480 B.C.) and his followers in southern Italy argued that the ultimate reality was not material but numerical.

Plato (427–348 B.C.), Socrates's most famous pupil in Athens, argued that there were two worlds—the world of forms and the material world. A form, he said, was an eternal, perfect, preexisting idea—a sort of blueprint—that was imperfectly replicated in the physical world, because matter had limitations and was transitory. Plato believed that the material world was constructed by a divine craftsman according to a formal plan. He explained his view in *The Republic* with the Allegory of the Cave: We are like prisoners chained inside a cave and only able to see shadows on the cave wall, he said. We can know nothing of the reality of beings or objects outside of the cave except what we might infer from their shadows. The shadows are the world of sense experience. Our goal is to escape the bondage of sense experience and climb out of the cave in order to gaze on eternal realities, the world of forms.

In *Timaeus,* Plato offered an extended account of a cosmos designed according to mathematical principles. He conceived the elements as pure number and shape; atoms were composed of triangles, squares, and pentagons arranged into regular mathematical solids, such as cubes. He then associated each solid with a traditional element—earth, air, fire, or water. Plato was offended that the pre-Socratics had deprived the world of divinity, plan, and purpose. They had gone too far when they had banished the interfering gods of Greek mythology. For him, an external, divine source had been required to establish cosmic order from a preexisting chaos. The cosmos was the product of reason, planning, and purpose, he believed, and could not have arisen from the chance arrangement of atoms or elements.

A MATHEMATICAL UNIVERSE

By the fifth century B.C. astronomical knowledge from Mesopotamia had been passed on to the Greeks. This included the division of the sky into the 12 signs of the zodiac and observations of the apparent motions of the stars and planets, such as the daily rising of all heavenly bodies in the east and their setting in the west; the eastward drift of the sun, moon, and planets in relation to the fixed stars; and the looping motion of planets, which moved east for weeks or months, sometimes slowed down, reversed direction, and headed west until they again reversed direction and continued on their original eastward course. This was known as the retrograde motion of the planets.

The fifth century B.C. was also marked by progress in geometry and mathematics and by the emergence of a belief in the value of geometric reasoning to solve problems of nature. In keeping with his view that the universe was fundamentally mathematical—that simple mathematical elements gave rise to the apparent complexity of the world—Plato proposed that the observed irregular and retrograde motions of the planets could be explained in terms of combinations of uniform circular motions. Two of his contemporaries, Eudoxus of Cnidus (ca 400–347 B.C.) and Aristotle (384–322 B.C.), took up this proposal. Both considered the Earth stationary and at the center of the universe. Surrounding the Earth, they argued, was a series of nested spheres centered on the Earth but rotating uniformly about different poles and at different speeds. In this model, the outermost sphere contained the fixed stars, and it made a complete rotation in 24 hours. Other spheres between the sphere of fixed stars and the Earth carried the moon, sun, and planets. The compounded rotations of these spheres gave rise to the observed motions of the celestial bodies, Eudoxus and Aristotle claimed.

18

Plato (left) believed the world was designed according to a plan that could be discerned through mathematics. His student Aristotle (right) believed observing nature was the key to understanding.

Aristotle was Plato's student, but he challenged his teacher on key issues. Whereas Plato distrusted the world of the senses, Aristotle, an experienced biologist and natural historian, took sensory experience as the starting point for his philosophy. Aristotle also contended that the forms of objects did not have prior or higher existence than matter. Corporeal things were made up of forms, such as color, weight, or a particular character, such as "dogness," united to and inseparable from the matter that served as the basis for the forms. When an object changed, its form also changed.

Aristotle's cosmos was a world of purpose and order, and not chance and coincidence. For Aristotle, the best way to study nature was to observe it in its natural, unfettered state.

Aristotle went further to expound a physical model to account for the observations. He believed that the cosmos was divided into two main parts—the celestial realm of the stars and planets, and the sublunary (below the moon) realm of the Earth and its atmosphere. The sublunary realm was characterized by birth, death, generation, growth, and corruption. All things below the moon were made of earth, air, fire, and water, or mixtures of these elements. Natural motion was motion toward a body's natural place. Air and fire tended to move upward, earth and water downward. When a body reached its natural place, its motion ceased, Aristotle observed. Motion in any other direction was called forced or violent motion. Violent motion required a motive force, a mover. For instance, if a heavy stone were thrown upward, its motion was violent and counter to its natural downward tendency. Once the impetus of the violent motion was exhausted, the stone, striving for its natural place, would fall to the Earth.

In the heavens, things were different. The heavens began with the moon and included the sun, planets, and stars. They also included transparent, celestial spheres that carried the celestial bodies in unending circles. The celestial realm was made of aether, the fifth element, which was eternal, unchanging, and quasi-divine. The natural motion of aether was believed to be circular.

Aristotle's cosmology required two different types of dynamics. Everything below the moon was governed by terrestrial physics. Everything above the moon was governed by celestial mechanics. The Earth was thought to be immobile. Because the terrestrial elements rose from or fell toward the center of the Earth, the Earth itself had to be resting in its natural place in the center of the universe. Because parts of the Earth could move only up or down, it was absurd to think of the whole Earth moving in a circle. The moon, Mercury, Venus, the sun, Mars, Jupiter, Saturn, and the fixed stars were carried (in this order) around the Earth by a system of nested, concentric, celestial spheres. There were 55 spheres for Aristotle. By compounding the motions of different spheres he was able to explain the retrograde motion of the planets. Each sphere was moved by an unmoved, spiritual mover or the sphere above it, but the ultimate cause of every celestial motion was the

19

Claudius Ptolemy's mathematically based model of an Earth-centered universe stood the test of time. It was accepted throughout much of the world for nearly 1,500 years.

Prime Mover, or Primum Mobile, who was a deity wholly separate from the world. The Prime Mover caused motion in the heavens not through contact with celestial bodies, but as the object of desire of these celestial spheres, which tried to imitate its changeless perfection by spinning uniformly forever. Aristotle's cosmos was hierarchical and organized according to degrees of perfection. Things near the center of the universe (the Earth) were more imperfect than those near the outer limit (God). Everything had its natural place and strived to reach that place.

Aristotle's physics and cosmology held center stage for nearly 2,000 years. There were notable challenges to the Aristotelian view, however. The Pythagoreans taught that the

Astrolabes are sophisticated, handheld instruments that were used for centuries as mechanical maps of the universe, as calculating and teaching devices, and as traveler's navigational aids. The concept of the astrolabe dates from classical Greece, where philosophers such as Hipparchus (180 B.C.) explored ways to project maps of the sky onto flat surfaces. The most well-known early discussion of an actual instrument using this form of sky projection is found in Claudius Ptolemy's Planisphaerium (ca A.D. 150). By the tenth century the Islamic world had adopted and refined the astrolabe.

Usually made of brass, astrolabes were used for centuries to teach people about the sky and to take observations of sky phenomena for astronomical, civil, and religious purposes. Holding the astrolabe by a small ring with one hand, the astronomer sighted along the movable pointer, called an "alidade," lining up the sun, the moon, a planet, or a bright star, and then carefully recording what the angular altitude of that object was above the horizon. This simple observation linked the astronomer's position on Earth, the time of day or night, and the date of the year with the place of that object in the sky. All the complex circles and charts on the astrolabe were mechanisms to calculate how these quantities were related and allowed the astronomer to use the sky as a great clock. Most often it was used as a powerful tool and as an authoritative symbol for the casting of horoscopes.

Astrolabes came in many sizes and styles, but their basic function did not change between the 10th and the 17th centuries,

either in the East or in the West, which demonstrates the relative stability of the science of astronomy during this period. One of the oldest astrolabes in the collection of the Smithsonian's National Museum of American History was constructed in Muslim Spain ca 1090 and is among the most ancient in existence. This four-inch radius astrolabe has several interchangeable plates, each engraved with the local coordinates for a different latitude. The pointers on the top plate indicate the positions of 22 bright stars. The top plate can rotate to show where those stars will appear at different times or dates, much like a modern paper or plastic star finder. The instrument could be used to predict when the sun or certain bright stars would rise or set on any date, but most often was probably used to tell the time of prayer. Eventually this astrolabe passed from Arabic hands into the West. As it did, many of its original Arabic inscriptions were probably erased to allow for Hebrew inscriptions. Both Arab Kufic and Hebrew inscriptions are still to be found on various parts of this extraordinary device. Another example, from 16th-century France, is a bit larger in diameter, but otherwise functions the same as the 11th-century example. Known as the "Galois 1548," it was constructed for use at a latitude of 48 degrees north.

The astrolabe was slowly introduced into Western culture through Muslim Spain. By the 13th century, Latin texts were available describing them, though the device gained its broadest popularity starting in the late 14th century, after Geoffrey Chaucer recommended it as an important astrological

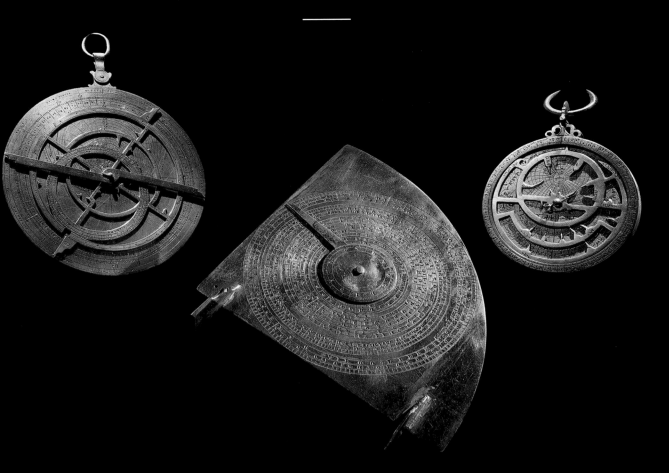

forecasting device. It became a central instrument in the education of the elite. Although European instrument makers discarded the Islamic prayers inscribed on earlier astrolabes, they retained the Arabic names of the stars. Some 1,600 astrolabes of all types survive today, though only a few hundred have been thoroughly studied.

Some astrolabes combined the features of a quadrant and an astrolabe. Only seven of these devices are known to exist today. This one, located in the center of the grouping above, dates from 1325. The circular face of this brass astrolabe has essentially been "folded over" twice to fit on a quarter-circle. It could serve as a measuring tool and perform many of an astrolabe's calculation func-

tions. It was used as a sighting device for astronomical and terrestrial observations.

Quadrants were generally easier to use than astrolabes and ranged in size from small handheld versions to large mural quadrants many feet across. After the invention of the telescope, large quadrants were often combined with telescopic sights for improved positional measurements. Today, the ancient technology of the astrolabe and the quadrant form the foundation for the planispheres and moving star charts that adorn telescope shops, planetarium bookstores, and public star parties, helping us maintain the connection between the Earth and sky that has existed for centuries.

—*David DeVorkin*

The French astrolabe (left) dates from the 1500s; the Islamic astrolabe (right) was crafted in 1090. The medieval instrument between them—only a few still exist— combines features of an astrolabe and a quadrant, a tool for measuring altitude.

Earth was neither stationary nor at the center of the universe, but moved around a central fire. Aristarchus of Samos (ca 310–230 B.C.) believed that the sun was the center of the cosmos and the Earth orbited it. The Stoics rejected the division between the celestial and earthly realms, arguing that a pervasive spirit united man with the universe. Followers of Epicurus (341–270 B.C.) believed that there were an infinite number of worlds in an infinite space. Natural philosophers of the 16th and 17th centuries, such as Copernicus, Descartes, and Newton, would later cite the arguments of these earlier philosophers as justifications for their radical views.

PTOLEMY'S UNIVERSE

Five hundred years after Aristotle, Claudius Ptolemy (ca A.D. 130–175) in Alexandria brought to planetary astronomy a level of mathematical sophistication unknown at the time of Aristotle. Like his predecessors, he wanted to discover some combination of uniform circular motion that would account for the observed positions of the planets, but his goal was to be able to make accurate predictions of planetary motion, which earlier models had failed to do, and to account for the changing brightness of the planets, which he presumed to be caused by their varying distances from the Earth. He therefore set aside the notion that the circular motions must be concentric to the Earth and used new mathematical models that were to predict the observed positions of the planets, or "save the phenomena." At first, Ptolemy did not care how physically implausible his new geometrical models might be. In his *Almagest*, he aimed for mathematical simplicity. In a later work, the *Planetary Hypotheses*, he argued, however, that the models he had presented in the *Almagest* had real, physical existence in the world.

Ptolemy's planetary model had three key components:

• **eccentric circles**—circles not centered on the Earth. When a planet moved uniformly around the eccentric circle, it appeared to Earthbound observers to vary its brightness and speed.

• **epicycles on deferents**—small circles (epicycles) whose centers moved along larger circles (deferents). An epicycle could carry a planet uniformly around a deferent but create the appearance of backward, or retrograde, motion.

• **equant**—a point in space about which the planet moves uniformly, producing the impression of nonuniform motion from the vantage point of the Earth.

Ptolemy's mathematical model of the heavens was combined with Aristotelian physics (though seldom seamlessly) to create a cosmology that was resoundingly endorsed by West-

A medieval artist depicted God as a geometer and architect, a view shared by many astronomers of the time. The quest to discover the geometrical plan that explained the organization of the universe obsessed many thinkers from the ancient Greeks through the Middle Ages.

ANCIENT COSMOLOGIES

ern astronomers for many centuries.

With the collapse of the Roman Empire, the Latin West lost vital contact with the Greek East and its great astronomical traditions. Knowledge of Greek achievements was limited to short descriptions included in Latin handbooks, encyclopedias, and commentaries. Primary sources, especially Greek ones, were lost.

Early Christian church apologists borrowed from Greek philosophy in developing church doctrine. One of the few ancient philosophical works known in the early Middle Ages, Plato's *Timaeus*, had useful lessons. Plato's cosmology emphasized divine order and guidance and could be read as a monotheistic response to pagan beliefs in multiple gods or atheistic materialism. Others regarded Greek philosophy—especially that of Aristotle and Epicurus—as a source of error. Tertullian (ca A.D. 155–230) saw it as a source of heresy. Less extreme, Augustine (A.D. 354–430) thought philosophy was useful, if not always reliable. Philosophy, he said, was to be the handmaiden of religion; the study of nature could contribute to the proper interpretation of the Bible.

THE MIDDLE AGES

The Byzantine Empire was more stable than the Roman Empire, and its capital at Constantinople did not fall until 1203. Intellectual activities declined there, too, but less rapidly than in the Latin West. More important, Hellenistic philosophy spread from Byzantium into Asia and North Africa. With the arrival of Islam in parts of Asia, North Africa, and the Middle East by A.D. 750, followers of Muhammad came into contact with Greek philosophy. Agents were sent into the Byzantine world in search of manuscripts to translate into Arabic. Of note was the House of Wisdom in Baghdad, a research institute founded by the Caliph al Ma'mūn (rule 813–833) and overseen by a Nestorian Christian, Hunayn ibn Ishāq (808–873). Hunayn, his son, his nephew, and others undertook the translation of many Greek and Syriac works into Arabic, including those by Aristotle, Plato, Euclid, and Ptolemy. By A.D. 1000 a large body of Greek natural philosophy, medicine, and mathematics had been translated into Arabic.

Muslim scientists extended the science they inherited. Their astronomers produced very sophisticated work, improving on Ptolemy, compiling planetary tables, and developing new instruments for research. Al-Farghāni (d. 861), a court astronomer in Baghdad, wrote a textbook on astronomy that circulated widely. Thābit ibn Qurra (d. 901), another court astronomer, studied the motions of the sun and moon, concluded that the westward motion of the equinoxes was nonuniform, and developed the theory called trepidation to account for it. Al-Battāni (d. 929) mathematically improved the Ptolemaic models, studied solar and lunar motions, and gave instructions for building astronomical instruments such as a mural quadrant, a large, fixed instrument that measured the altitudes of celestial objects.

Back in Europe, scholars heard rumors of the scientific riches preserved in the Muslim world and went in search of them, especially in Muslim Spain, at the end of the 10th century. Translation became a major scholarly activity. By the late 12th century original Latin and

Greek works, as well as more recent Arabic treatises and scientific instruments, had been recovered. Technical treatises such as those of Euclid and Ptolemy were peacefully received and were often taught at the universities. Aristotelian cosmology, however, raised some eyebrows. On a number of occasions in the 13th century, the bishop of Paris condemned Aristotle's natural philosophy for being deterministic and pagan, denying the immortality of the soul, and asserting the eternity of the world. Nevertheless, through the work of theologians such as Albertus Magnus (ca 1200–1280) and his pupil, Thomas Aquinas (ca 1224–1274), the Aristotelian cosmology was Christianized and made an acceptable and robust handmaiden to theology. For instance, Aristotle's spiritual movers were replaced by angels, the Prime Mover was taken to be God, and the outermost sphere of the cosmos became the habitat for the elect. In this way, reason and faith were harmonized.

With the entrenchment of Aristotelian cosmology in Western intellectual circles from the 14th through the 17th centuries, new tensions arose in efforts to reconcile the Ptolemaic planetary model with Aristotelian physics. If heavenly matter, or aether, moved in eternal circles, why were the planetary motions apparently not uniform or centered on the Earth, as Aristotle claimed? If Ptolemy was right, would not the planetary mechanisms of overlapping epicycles and deferents get in each other's way? Even more troubling was the equant. What physical reason was there for a planet to revolve around this imaginary point? One solution was to argue that these devices must be mathematical fictions to save the phenomena and not real mechanisms in the heavens. Another tack was taken by Georg Peurbach (ca 1423–1461) who proposed a materialized construction of the Ptolemaic geometrical devices. According to this world system, each planet was embedded in an epicycle, or small circular movement, that rolled between two adjacent circles not centered on the Earth (eccentrics), and all components of the celestial spheres were made of a crystalline substance.

Such reconciliations did not satisfy everyone, but as a practical matter, many astronomers were not troubled by these inconsistencies. Others like Copernicus were led to consider major reforms to cosmology.

EARLY ASTRONOMICAL INSTRUMENTS

The debate over the geometric and computational models of the cosmos was carried out not only in lively discussion and written tracts, but also with scientific instruments. This part of the story has always held particular fascination for me. While working in the collections at Harvard, the Adler Planetarium and Astronomy Museum, and elsewhere, I began to examine preserved records and surviving instruments, hoping to discover who designed and produced the instruments, how they were used and by whom, what functions they served, and how their forms not only followed their functions, but also how they could embody the scientific theories that they were used to explain. I also began to make replicas of historical instruments in order to make observations in the old ways. These hands-on activities helped me to understand the sophistication of early astronomy and the challenges its practitioners faced.

Deceptively simple, the gnomon (a shadow-casting stick) was a sophisticated astro-

Astronomers at work. Western astronomers observe the heavens to discern horoscopes in a 14th-century manuscript (above). Islamic astronomers at the observatory of Taqf ad-Din at Istanbul in 1577 (right) use instruments, including an astrolabe, quadrants, a tripod, a long-sighting device called a parallactic ruler, perhaps a small armillary sphere, a mechanical clock, and a globe of the Eastern Hemisphere.

ANCIENT COSMOLOGIES

nomical instrument for its time. Anaximander of Miletus set up a gnomon in Sparta in the sixth century B.C. By tracing the lengths and angles of the gnomon's shadow, Greek philosophers could mark the times of the solstices and equinoxes, measure the number of days between them, establish calendars, and project the celestial sphere onto a flat or curved surface. The combination of the gnomon and the mathematical projection created the sundial. By the third century B.C. hour-finding sundials were common in Greece and Rome. They shed light on the place of mathematics and astronomy in people's daily lives.

The celestial globe was another projection of the heavenly sphere. Early globes documented the positions of major star groups, along with the principal celestial circles—the Equator, the tropics of Cancer and Capricorn, the Arctic and Antarctic Circles, and the ecliptic (the path of the sun against the fixed stars). These globes were used to solve basic problems of positional astronomy. Eudoxus of Cnidus possessed and used a globe, as did Ptolemy five centuries later.

Globes were teaching tools as early as the third century B.C. in Greece. Globes embodied the essential astronomical and cultural cosmology of the day. They represented the universe as spherical and closed, as if it were being viewed by a spectator beyond the universe (i.e., from the vantage point of God). Moreover, globes delineated the constellations, which were used both as scientific maps and allusions to the traditional gods of Greek mythology. This rich combination made globes prized scientific instruments as well as cultural symbols of power and prestige. The earliest surviving globe is the one borne on the shoulders of the Farnese Atlas, a Roman statue dating from about 200 B.C. and preserved in the Museo Archeologico Nazionale in Naples.

Another model of the universe and a close relative to the celestial globe was the armillary sphere, composed of rings representing the great circles of the celestial sphere. Two forms of the armillary sphere descended from antiquity; one was an instrument for observation and the other was a tool for teaching. The observational instrument was devised by Ptolemy to measure the positions of stars and planets in the coordinate system of his choice. Observational armillaries were a standard fixture in medieval Islamic observatories. Information on the construction and use of this instrument was transmitted to the West in the late 12th century. Nicolaus Copernicus was among those who used an observing armillary sphere in the early 16th century. Tycho Brahe refined its design and use at the end of the century.

The teaching armillary sphere was a scaled-down version of the larger observational instrument. Between the outer rings representing the sphere of fixed stars and a central globe representing the Earth, this instrument had a series of nested, movable rings or spheres for the moon, the sun, and the planets. Thus, the armillary embodied the Earth-centered cosmology of Aristotle and represented our place in the universe. It was a standard piece of equipment for astronomers and scholars from the medieval period through the 18th century.

The quadrant was another important angle-measuring instrument for the astronomer. The plinth, a forerunner described by Ptolemy, used a peg to cast a shadow on a vertically standing 90-degree arc to measure the altitude of the sun. Al-Battānī, the Muslim astronomer, had a large quadrant fixed to a north-south wall in his observatory. He used an alidade, or sighting rule, to make his observations. Nasīr al-Dīn al-Tusīn Maragha had movable quadrants that could measure angles of altitude or azimuth (the bearing of an object). These quadrants were adopted by western astronomers during the Renaissance. Small portable quadrants were used in medieval times for surveying and determining time.

The queen of ancient scientific instruments, however, was the astrolabe, which simulated the apparent rotation of the stars around the celestial North Pole. It consisted of a "see-through" star map that rotated upon a stack of engraved plates, each representing the sky as seen from a different latitude. Once the user selected the plate for his particular latitude, he could determine where and when the celestial bodies would move in the sky from his vantage point. The astrolabe was used for locating stars in the sky and finding the times of their risings and settings; for determining the hour, day or night; for making astrological calculations; for finding latitude or longitude; for surveying; and, in Islamic regions, for trigonometric calculations and determining the hours of prayer as well as the direction of Mecca. In short, the astrolabe was an analog computer and portable model of the heavens. Developed near Alexandria before the fourth century A.D., it was widely used and improved in the Islamic world before arriving in the Latin West through Spain in the tenth century.

To the inventors of the astrolabe and the quadrant, to the brilliant minds who saw how a simple stick could teach them about the movement of the sun, to the philosophers and mathematicians, and the peoples of the Earth who found purpose and order in their observations of the night sky, the scientists of today owe a great debt. Our appreciation and understanding of the universe has been built on these culturally rich foundations.

RIGHT: Astrolabes such as this Islamic one were used to calculate the time of night by observing the altitude of bright stars indicated by the pointers on the top plate.
LEFT: Among the earliest astronomical instruments, sundials were common time reckoners in Greece and Rome by the third century B.C. This Roman sundial, made of marble, dates from the first or second century A.D.

EAST ASIAN
COSMOLOGIES

John S. Major

In 1669–74 the Belgian Jesuit priest Ferdinand Verbiest carried out a commission from the Kangxi Emperor of China to create astronomical instruments for a new imperial observatory in Beijing. Verbiest, an accomplished astronomer, performed his work well, and his instruments can still be seen atop the Old Observatory Tower, east of Tiananmen Square. The instruments, cast in bronze with decorations of clouds and dragons, look typically Chinese at first glance, but closer inspection reveals that they are almost entirely European in inspiration. They include such characteristically early modern European instruments as a quadrant, a sextant, and a celestial globe marked with the European constellations. Although Verbiest worked several decades after the discoveries of Galileo, the taint of heresy that still clung to

Belgian priest Ferdinand Verbiest crafted astronomical instruments for Beijing's new imperial observatory in the mid-1600s. As Verbiest and other Jesuit astronomers introduced European cosmological theories to China, an astronomical tradition that spanned two millenia began to fade.

南公懷仁像

those discoveries is probably why the Beijing observatory did not include a telescope. Even so, the overall impact of Verbiest's instruments, and the European astronomical theories introduced along with those instruments by the Jesuit astronomers in 17th-century China, marked the beginning of the end of a 4,000-year tradition of indigenous Chinese astronomical theory and practice.

Human figures personify the 28 equatorial constellations, or lodges. As in the West, astronomers used star groups to track the movement of the sun, moon, and planets.

The beginning of the story is lost in the depths of time, but it is safe to assume that the ancient ancestors of the Chinese studied the night sky, named stars and constellations, tracked the movements of the planets, recorded the phases of the moon, wondered at such irregular events as the appearance of comets and meteors, and devised myths to explain what they saw. Not everyone in the scholarly world would agree with me, but I would argue that China had astronomers before it had kings.

More than 4,000 years ago, near the dawn of the Bronze Age in East Asia, astronomy began to change from folklore to a science. As reference points to help them more easily track the movements of the moon, the planets, and the sun when it was below the horizon, astronomers selected and named a group of 28 constellations (called *xiu,* or "lodges") arrayed along the celestial equator. These lodges are similar but unrelated to the 12 houses of the Babylonian zodiac. The number 28 was probably chosen because Saturn, which has the longest orbital period among the visible planets, takes approximately 28 years to make a complete circuit of the heavens. The system of using lodges as reference points apparently spread at an early date to India (where the lodges became known as *nakshatras*) and from there to the Arab world.

At about the same time, the Chinese devised the basic elements of their calendar, which is one of the oldest continuous calendrical systems in the world. Every calendar that tracks both solar and lunar time must reconcile two awkward numbers: the 365.25 days of the solar year and the approximately 354 days of 12 complete lunar cycles, or months. The solution lies in adding extra months—seven per 19 years, or slightly less than one every three years—to keep the two cycles roughly in balance. In later eras, as calendar makers became more sophisticated, their calendars were capable of running for decades or centuries without needing to be adjusted.

The invention of the calendar was ascribed by later Chinese historians to the Yellow Emperor, a mythical figure who supposedly ruled for a hundred years during the third millennium B.C.; he can be thought of as a personification of the cultural inventiveness that characterized that era, in the high Neolithic and the early Bronze Age.

Around that same time the Chinese began to visualize the sky as a hemisphere divided into five "palaces"—the Purple Palace of the Center, the disk of permanently visible circumpolar stars, which never dip below the horizon, and the palaces of the Bluegreen Dragon of the East, the Vermilion Bird of the South, the White Tiger of the West, and the Dark Warrior of the North, each a swath of sky stretching from the circumpolar disk to the celestial equator and centered on its cardinal direction.

Finally, the pre-dynastic period saw the invention of the Chinese day-count system, which combines two sets of ordinals—the 10 Heavenly Stems and the 12 Earthly Branches. If the Heavenly Stems were the letters A through J in our alphabet, and the numbers 1 through 12 were the Early Branches, the way the Chinese combined them would look like this: A1, B2, C3, D4, E5, F6, G7, H8, I9, J10, A11, B12, C1, D2, E3, and so on. This produces a repeating series of 60 day enumerators (half of the potential 120 combinations) that are used to name and count days. (Much later, the same counting system was applied to create "centuries" of 60 years each.) The day-count system apparently arose because, with the increasing complexity of late Neolithic and Bronze Age civilization, it was important to keep track of days so as to carry out ancestral sacrifices and other ritual obligations in a timely fashion. The first Chinese royal dynasties—the Xia (ca 1950–1550

Emperor Fu Hsi holds a yin-yang, a powerful symbol of the complementary forces in the universe: yin, the female, symbol of Earth and dark; yang, the male, of light and heaven.

B.C.) and the Shang (ca 1550–1045 B.C.)—were obsessive time recorders; Shang divination records, carved into turtle shells and animal bones, already used the dating system that was to endure for centuries:

"N" year of King so-and-so, month "X," day "Y."

The Chinese never had an index year—a "year one"—in their calendar. Today, on Chinese restaurant place mats, for example, one sometimes sees a date such as "Chinese year 4688," but that is simply a retrospective calculation of years back to the supposed reign of the Yellow Emperor. This cumulative year count was never used in China until very recent times, although now it has begun to appear occasionally in Chinese publications. (Among other things, it is flattering to Chinese cultural chauvinism to have such a long year-count.)

As far as we can tell, no great advances in astronomy or calendrical science were made during the late Shang dynasty and the early centuries of the Zhou dynasty (ca 1045–256 B.C.). But in the latter part of the Zhou dynasty, known as the Warring States Period (481–221 B.C.) because of the ferocity of its internecine strife, the astronomical sciences shared fully in the burst of creative energy that marked those turbulent times. A significant expansion of literacy led to an explosion of writing and copying texts. Philosophers, trying to understand the sources of the warfare and treachery that marked the politics of their era, systematized thought in China and led the way from a mythico-religious worldview to a more rationally based one. Astronomers created star maps and catalogs of star names, improved their skill at calculating lunar and solar eclipses, and refined their estimates of the orbital periods of the planets. Given the prevailing view that the Earth was flat and square, and the heavens round and domed, some asked how the sun could rise, cross the sky, set, and travel back beneath the Earth. Using calculations based on shadow-line measurements, others answered that it was all an optical illusion based on China's distance from the North Pole. Astrologers calculated the correspondences between political states on Earth and sectors of the sky with their associated stars and constellations, and used the turning pointer of the Northern Dipper (or, as we would call it, the Big Dipper) to predict which states would win or lose in battle, which would prosper, and which would fail.

The Qin (221–206 B.C.) and Western Han (206–7 B.C.) dynasties followed the Warring States to create a unified empire from the discord of earlier times. In this period, artisans began to cast bronze mirrors, the backs of which were decorated with microcosmic patterns. It now became possible, in effect, for the possessor of such a magical implement to hold the universe in his or her hand.

At the same time, the astronomer/astrologer (one and the same at the time) invented China's first complex astronomical instrument, the *shi*, or *shipan*. The instrument's name is often translated as "diviner's board," but I prefer astromythology scholar Stephen Field's inspired translation of "cosmograph." By turning the heaven-plate of the cosmograph daily, in imitation of the motion of the heavens, astrologers could "follow the strike of the Dipper" and make their predictions without having to look at the sky.

My own involvement in the adventure of Chinese cosmology began with the Western Han dynasty. I was fortunate enough as an undergraduate to have Professor Derk Bodde as my mentor. Bodde is one of the last surviving members of the remarkable generation of Western sinologists who studied the language, literature, history, and culture of China in Beijing in the 1920s and 1930s. His tutors were scholars who had served under China's last emperors, and who were able to experience Confucian tradition as a living culture. Professor Bodde allowed me to write a senior thesis on the Han scholar-politician Dong Zhongshu, even though (as I had begun studying Chinese only two years before) I was unable to read classical Chinese texts in the original, and had to depend on translations of my main sources. That initiation into the world of ancient Chinese was enough, though. I had found my career.

Going on for graduate work at Harvard, I found a second mentor, Professor Nathan Sivin (then of MIT, now at the University of Pennsylvania), who encouraged me to register for his History of Chinese Science seminar. It was he, too, who suggested that for my seminar paper I should translate part of the *Huainanzi*, an eclectic work of philosophy, science, and statecraft completed around 139 B.C. under the direction of a philosopher-king named Liu An. (I like to describe the *Huainanzi* as a handbook of everything the modern monarch of the second century B.C. needs to know.) That relatively short translated passage from the *Huainanzi*'s "Treatise on Topography" became the nucleus of my doctoral dissertation, and was the beginning of a career-long involvement with Liu An's compendium.

As I pursued the next stage in my journey of discovery "dissertation-into-book" (which, after many years, resulted in the publication of *Heaven and Earth in Early Han Thought*), it became clear to me that the project would not be complete unless I also translated two other *Huainanzi* treatises: the cosmological "Treatise on the Patterns of Heaven" and the calendrical "Treatise on the Seasonal Rules." As I prepared to do so, I boldly, perhaps brashly, invited myself to Cambridge University in the spring of 1972 to visit the famous and venerable Joseph Needham, who, in his multivolume *Science and Civilisation in China*, had virtually invented the field of the history of Chinese science. Needham, famous for his courtesy and generosity, received me as a colleague rather than as the neophyte I really was; I left Cambridge a week later feeling that I had gained a friend for life, as indeed I had. Tucked away in my suitcase was a gift from Needham, a draft translation of the "Treatise on the Patterns of Heaven" done in the 1930s by an English engineer in China, Herbert Chatley. It was a very rough draft, but it at least gave me confidence that the almost impenetrably difficult language of the treatise could be forced to submit to the discipline of translation.

Later that same year I went to Paris to attend the International Congress of Orientalists, and there had an experience that I have cherished ever since. Taking a break from lectures, I wandered into the hall and noticed Wing-tsit Chan, the century's greatest historian of Confucian philosophy, standing alone. Having met and corresponded with him previously, I began to chat with him. Then I noticed Needham standing a few feet away, and realized after a few moments that he did not recognize Professor Chan. So I, as a very young scholar, had the immense pleasure of introducing two of the giants of Chinese studies; I left them happily conversing.

Later that day, Needham sought me out. He told me he had some materials on the history of Korean astronomy that needed "only a bit of editing" to make a book and asked whether I would be interested in preparing such a book as his co-author. I jumped at the chance. (Several months later I received three large cardboard cartons filled with an appalling jumble of notes, photographs, draft translations, and correspondence, and my heart sank. Eventually all of that did become a book, but not until many more years had passed.)

Meanwhile there was much work to be done on the *Huainanzi*. As a way to understand Chinese cosmology as a system that had made sense to some very learned and intelligent people 2,000 years ago, my scholarly approach was to not only work with the texts, but also to try to think my way into the mind-set of the men who had written them.

By the late 1970s I had come to three conclusions that have guided my work ever since. The first is that although the formal articulation of Chinese ideas about the universe did not take place until the late Warring States and into the Han periods, those ideas drew on a very ancient mythic tradition and formed a continuity with it. The second is that Chinese cos-

THE CONSTRUCTION AND USE OF THE COSMOGRAPH

The shipan, or cosmograph, is not unlike a simple Western movable star chart or planisphere relating the positions and arrangements of stars to the local sky of an observer at any time of day or night. Samples have been recovered from Han dynasty sites as old as 200 B.C. The cosmograph typically had a rotating circular disk (the yang or male component) representing the sky above a square base plate (the yin or female component) representing the Earth. There are numerous variants in form and interpretation; nonworking models were sometimes made as funerary pieces to be buried with the deceased, and not intended for practical use.

The edge of the moving disk is marked with the 28 constellations ("lodges") of the Chinese zodiac and the square base plate has concentric bands showing the same constellations, the names of the 12 Chinese months, and sometimes other indicators as well. In the center of the moving disk is the constellation known in the West as the Big Dipper, which the Chinese believed marked the throne of Shang Di, the "High God." Moving the disk allows the Dipper on the cosmograph to rotate as the real Dipper does in the sky. The motion of the disk on the cosmograph tracks the daily paths of these stars.

One uses the cosmograph much like a modern planisphere, turning the dial until the time and date are right to view the sky for any time of day or night and at any time of year. The significance of this dialing, to the Chinese, is based upon the connection of the 28 zodiacal constellations to regions and properties of Earth. Time, Earth, space, and the observer are all linked, making the cos-mograph a valuable predictive device for astrological forecasting. An observer would see a star grouping in the sky, near the south point. The cosmograph could tell what constellation was high in the sky above the south point at any other time of day, such as noon or midnight, astrologically auspicious times. It could also be used to calculate or "predict" which constellations would be rising at sunset or sunrise, or culminating (crossing over the north-south meridian line) at noon or midnight, at any time in the future.

Finally, like the "nocturnal," a simple Western device for telling the time of night from the orientation of the Big Dipper's handle, the cosmograph revealed the direction of the handle of the Dipper at any time. Because celestial locations were believed to have earthly counterparts, the cosmograph was thought to be able to predict good or bad fortune in particular places within the empire.

Until recently, scholars could only speculate about what the cosmograph looked like. Ancient texts described its function but not its form; only a few badly damaged fragments were known from archaeological excavations. Since the 1970s, several complete examples have been found in tombs, transforming our understanding of early Chinese astronomy and astrology.

The design of the cosmograph was related to other contemporary objects, including game boards and bronze mirrors decorated with microcosmographic imagery. One such mirror is illustrated here; its design depicts the square Earth within the circle of heaven, surrounded by the cosmic ocean.

—*John Major and David DeVorkin*

The cosmograph was one of several Han dynasty devices that embodied the concept of a round heaven and square Earth. The most beautiful of the Han microcosms are the so-called TLV mirrors, such as this one from the first century A.D. The mirrors (polished to a high gloss on the obverse side) were luxury goods, but also auspicious objects filled with cosmic power.

mology, and particularly Chinese astrology, rested on a concept of the cosmos that reduced it to a set of schematic figures, most importantly the ninefold divisions of heaven and Earth, represented as a grid of nine equal squares. The third realization I had was that Han cosmology and astrology could not be understood without reference to the *shi* cosmograph.

In Chinese formal cosmology of this period, space and time were integrated through the device of the grid of the ninefold division of Earth. Seasonal time rotated through the nine divisions governed by the waxing and waning of yin (dark, moist, female, etc.) and yang (bright, dry, male, etc.), the fundamental expression of cosmic dualism. Interactions among phenomena were mediated by a resonating, energetic "ether" called *qi*. Numbers, colors, musical notes, political territories, and much more were included in this grid symbolism, all of which was governed by time, symbolized by the annual sweep around the sky of the Northern Dipper, and the approximately 12-year orbital period of Jupiter. All of this information was encapsulated in the design of the cosmograph. In fact, the "Treatise on the Patterns of Heaven" is, in large part, a handbook for the operation and interpretation of the cosmograph.

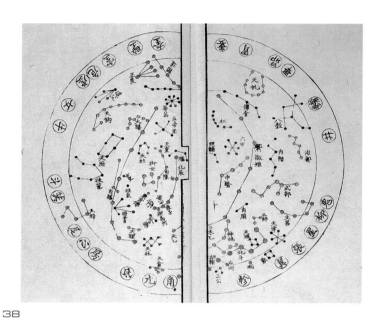

38

Chinese astronomers divided the sky into nearly 300 star groups. Twenty-eight lunar lodges, depicted along the border of this 15th-century chart, marked the daily passage of the moon across the sky.

But the cosmograph could not emulate the shape of the cosmos exactly. For the markings on the square Earth-plate to be visible around the rim of the round (and rotatable) heaven-plate, the heaven-plate had to be the smaller of the two. This obviously violated the old notion of heaven as a half-sphere completely enclosing a flat, square earth. In fact, the design of the cosmograph began to cast doubt upon the old domed-heaven paradigm. The Eastern Han (A.D. 25–220) philosopher Wang Chong (27–97) ridiculed the idea of a hemispherical heaven and a square Earth, commenting that "the corners would not fit," as indeed, in the cosmograph, they do not and cannot.

Partly for this reason, the extraordinarily learned Zhang Heng (78–139) of the Eastern Han invented China's first armillary sphere, an instrument embodying a quite different model of the universe. (Coincidentally, Ptolemy was making armillary spheres in Alexandria at about the same time.) Zhang's armillary continued to represent the Earth as flat, but surrounded by a spherical cosmos. The instrument's rings included the horizon, the prime meridian (where longitude would be zero), the celestial equator, and the ecliptic (the sun's path). Those are the basic rings of an armillary sphere; over the centuries that followed, other instrument makers added further features, such as a ring for the orbital path of the moon (near, but not on, the ecliptic), rings passing through the celestial poles that identifed the positions of the celestial bodies when the days were longest and shortest (the solstices) and the times when night and day were of equal length (the equinoxes), and sighting tubes that turned the instruments into observational devices as well as cosmological models.

In the centuries from the Han dynasty through the glorious Tang dynasty (618–907), Chinese cosmology, astronomy, and astrology made their way to the nearby countries of Korea and Japan. Korea's famous Ch'omsongdae observatory tower, constructed in 647, attests to this, as do various star maps and armillary spheres discovered in Japan that date from the 8th to the 11th centuries. In Nara, the Japanese capital from 710 to 785, the terrace on which the royal palace was constructed was called the Hokudodai, or Northern Dipper Terrace, showing that Japan's rulers, like the Chinese emperors, saw their throne as lying symbolically beneath the circumpolar stars, which never left the sky.

The armillary sphere models the cosmos nicely, but a static instrument still falls short of being able to display the dynamic motions of the heavens. Astronomers and instrument makers in the Tang dynasty, and into the succeeding Northern Song dynasty (960–1127), made repeated efforts to mechanize armillary spheres using water power and

This ingenious 11th-century clock tower was powered by water. Gears rotated a celestial globe and armillary sphere, imitating the apparent motions of the heavens.

gearwork to make them rotate in imitation of the actual motions of the heavens. After a number of more-or-less fruitful early attempts, these efforts reached their culmination in the work of the scientist-statesman Su Song (1020–1101), who, between 1086 and 1090, built in the capital city of Kaifeng an enormous, ingenious celestial clock powered by a waterwheel measuring 11 feet across that used gearwork and an escapement mechanism that paced the movement of the gears to faithfully reflect the passage of time. The clock featured a celestial globe and an armillary sphere, both rotating in imitation of the heavens, as well as jackwork puppets that appeared in a window of the clocktower to announce the time. Sadly, this magnificent instrument, as much a symbol of imperial prestige as it was a working astronomical tool, was destroyed in the sack of Kaifeng that put an end to the Northern Song period and forced the Song court to flee southward to Hangzhou.

The Southern Song fell to the Mongols in 1260 and, under the reign of the Mongol emperor Kublai Khan, Chinese astronomical instrument making saw its last period of greatness. Guo Shoujing (1231–1314), a mathematician and astronomer, made a new set of instruments for the imperial observatory. These included a new celestial globe, armillary spheres, sundials, gnomons, and a new device, the "simplified instrument," which was a dissected partial armillary with a sighting-tube, cast in bronze. Observers using the device could make highly accurate observations of the positions of heavenly bodies. Guo's instruments can be seen today at the Purple Mountain Observatory in Nanjing. Kublai Khan was also said by contemporary writers to have commissioned several automated water-powered clocks for the imperial palace, but no trace of them survives.

These water clocks and other medieval instruments formed the background for that book on Korean astronomy that I eventually did write with Joseph Needham, *The Hall of Heavenly Records*. Having worked for several years to bring order to Needham's cartons of jumbled notes, I returned to Cambridge for a year as a Visiting Fellow to work with Needham and our other colleagues. Our goal was to straighten out the history of the re-equipping of the Korean royal observatory in the 1430s under the aegis of Korea's greatest monarch, King Sejong (r. 1418–1450). We knew that King Sejong's observatory included a mechanized astronomical clock, and we fully expected it to be a waterwheel device similar to that of Su Song several centuries before. But, as we studied the texts and began to draw up working diagrams, it became clear that the waterwheel mechanism just would not fit what the texts were describing. In fact, as we eventually discovered, the Korean clocks were probably based on those of Kublai Khan, which in turn must have been powered by Arab-style inflow water clocks, or "clepsydras," rather than by waterwheels. In these devices, waterpower caused rods to rise by flotation, and the rods then released balls that rolled down chutes to power various clockwork effects. Here was an astonishing demonstration of the worldwide spread of knowledge under the Mongol Empire, with the ball-operated water clock chimes of Damascus influencing the design of an astronomical clock in Seoul.

Another instrument that we examined in that book is a 17th-century clock, now in the collection of the Korean National University, which also demonstrates the diffusion of ideas over long distances. It has a European-style weight-driven clockwork gear train, a Sino-Japanese-style time-announcing system, and a mechanized Chinese-style armillary sphere surrounding a terrestrial globe of purely European design. This clock was made in 1669, virtually at the same time that Father Verbiest was casting his bronze instruments in Beijing. In studying these events, we realized that the dawn of our modern era has not been solely a matter of the spread of Western thinking, but involved the flowing together of many streams of world culture; the Western stream has been dominant, but the waters mingle. It has been the privilege of my own life to savor some of the most exotic and long-lived drafts of these waters.

Many of the ornate astronomical instruments crafted in the 1600s by Ferdinand Verbiest for the imperial observatory in Beijing can still be seen atop the Old Observatory Tower, east of Tiananmen Square. In the foreground is a lavish equatorial armillary reminiscent of Tycho Brahe's instruments.

EAST ASIAN COSMOLOGIES

COSMOLOGIES OF THE AMERICAS

Von Del Chamberlain

About 30 years ago while visiting Colorado's Mesa Verde National Park I was reminded that there is no more awe-inspiring sight than a dark sky sparkling with thousands of stars. After years of using planetariums to introduce people to the wonders of the universe, seeing the cosmos under the night sky of this place so far from city lights inspired me to think about the astronomical awareness of those who built the impressive houses inside giant rock shelters along this mesa. This was the home of the Pueblo, and here their rock art and the walls of their elaborate cliff dwellings remain. I wondered what the people who lived in this place thought about the sun that warmed them during the day, and the moon, planets, and stars they would have seen so well on the clear, spectacular nights of this desert land.

Jupiter and Saturn gleam in the star-washed sky over the Anza-Borrego Desert in southern California. Native cultures throughout the Americas, especially in the harsh desert terrain of the Southwest, depended on an intimate knowledge of the sky for survival. Their cosmic mythologies are rich and diverse.

This question sent me to the library to find out what the Pueblo peoples of Mesa Verde and other Native Americans had known about astronomy several hundred years ago. I discovered that America's earliest inhabitants possessed sophisticated observational knowledge of the heavens, which they used to keep calendars, to move about the country, and to feel at home in the universe. I found story after story about the origins of things—Native American cosmologies that combined observations of nature with human experience and creativity. Clearly, the astronomical traditions born right here in America were just as provocative and interesting as any found elsewhere in all the world.

Native Americans, like societies around the globe, created explanations of origins, both of the things around them and of themselves. These mythologies tell its people who they are, how they came to be, how they are related to the things around them, and perhaps how they are to act while on the Earth and what will happen to them when they leave it. Even today, many of us seek a cosmic mythology—a human explanation for the origins of everything we find on the Earth and in the sky and our place in the universe.

Unlike modern science, which tends to put things into compartments—biology, geology, and astronomy—Native Americans tended to view all things as being interconnected. Sun, moon, planets, and stars might be all wound up with clouds, air, plants, animals, rocks, and people.

The complete picture of any culture's cosmic mythology can be elusive, however. Each is a complex tapestry woven from the threads of belief systems, stories, rituals, language, arts, architecture, and calendar systems. The inspirations for these designs—the natural phenomena taking place both on the nearby, touchable Earth and in the distant, unreachable heavens—are sometimes difficult to recognize. They might be encoded in ways that are not discernible to those outside the culture itself. Fortunately, this can often be at least partially reconstructed from its stories, ceremonies, symbols, and practices, some of which might even be expressed in the architecture of its buildings or the layouts of its cities, but it does not usually lie on the surface, neatly displayed in a museum, with pithy, authoritative labels.

For the Pueblo people, who lived on the high, dry plateau of the American Southwest, an intimate knowledge of the seasons was necessary for survival. Relationships between the sun, the Earth, moisture, and the passage of time were recorded and remembered in distinct ways. Creation myths were especially important mechanisms of remembrance. Each of the Pueblo groups has its own version of how the first people came into existence, but they all share the idea that they somehow emerged from Mother Earth to live in the spectacular open high desert country. According to Hopi cosmology:

> At first they were like ants, living in the womb of Mother Earth. Moving to another level they turned into other creatures, and still higher they became more like people, but having long tails. After a time a flood was coming and they sought a way to reach the sky. They planted a pine tree, but it would not reach high enough. Finally they planted reeds which grew tall to

penetrate the sky. They sent Badger up, but it was dark and he could not see much. Then Shrike went up and saw fire off in the northeast. Going over there Shrike found Masauwü [the fire god] who told Shrike to go back down and bring the people out.

They came out onto the surface of Mother Earth just in time to escape the flood. Mocking Bird gave them languages [Hopi, Navajo, Paiute, Shoshone]. Still, it was dark so the

Warrior Twins went to a place and began putting up stars in proper patterns. They placed the Seven [Pleiades], the Six [Orion], the Dipper and other stars. Thinking this work would never end, Coyote grabbed the remaining stars and scattered them in every direction.

Native Americans living along the Pacific Northwest coast told part of their creation myth through the stories of Raven. These peoples depended upon the sea as well as the land, and it played a large part in their cosmology. Many tribes in that part of the country tell stories about Raven, who was both creator and trickster. These stories say that the world would not be the way it is without this curious creature. Raven was also human, indeed superhuman. He was always exploring, finding interesting things to do and ways to change the way things are. Creation myths about Raven showed

45

Astronomical symbols, such as the moon and star pictured above, often appeared as pictographs on rock walls. This one in Chaco Culture National Historical Park, New Mexico, might represent the supernova of A.D.1054.

how sunlight could make things grow and provide food for Earth's family of creatures, how moonlight diminished the fearful darkness of night, how stars gave texture to the sky, tracing pathways, defining directions, and creating questions in our minds that result in voyages throughout the Earth and beyond into the grand cosmos. Water, substance of life, prime ingredient of this planet, played an important role. Having improved the world, Raven came upon something new and curious:

Now that there was light, Raven began to explore a spacious new world ... he flew to the seashore, landing on a sandy beach. Walking along, talking to himself as ravens always have, he checked out colorful pebbles, shapely shells, and nibbled morsels brought in by the charging waves. In and out, he dodged the surf, playing the seashore game.

So involved was he with all the tiny things to investigate, that he nearly stumbled over a gigantic clam shell that had washed up from the sea. Raven walked around this very curious thing, looking at it from every angle. It was most interesting. Not only was it large, but all along the sides of the shell there were peculiar, wiggling things sticking out. Strange things inside were trying to escape.

46

High on the face of a cliff on Fajada Butte at the south entrance to Chaco Canyon, New Mexico, Anasazi artisans carved a spiral petroglyph (above) that signals the start of each season. Slabs of rock create daggers of sunlight that move across the petroglyph to mark the solstices and equinoxes. A single shaft of light bisecting the spiral at midday marks the summer solstice. Rising dramatically from the flatness of Chaco Canyon, Fajada Butte (right) was likely considered a sacred site by the Anasazi. The rock art created there has been studied by a wide range of researchers—from geographers to ethnographers.

Raven sat and watched for a while, wondering what those squirming things were inside this giant shell. Perhaps he could play with them, or maybe they might be good to eat. Finally, curiosity got the best of him, as it always did, and he invented a way to release the creatures from their prison. They jumped onto the ground, running all about, laughing and speaking a language that Raven would have to learn. Instead of remaining to play, they ran off into the forests and hills.

Raven had released the first humans, who had come up out of the sea in the great clam shell. A major change began to take place on that day when he let those first people out onto the land.

The geographical and physical setting, the histories of the Northwest Coast people, the inherently intertwined natural phenomena, and the human characters all emerge as elements of a story that is compelling, memorable, and authoritative.

The Hopi, in contrast, were influenced not so much by the primacy of water as by the magnitude of the landforms that defined their universe. Deep gorges, huge sculpted rocks, and other dramatic landforms had to be rationalized somehow. Thus the Hopi explained that in the early days of formation,

. . . the earth was soft and muddy from the flood. The Twins shot lightning arrows into the ground, making canyons, hardened mud into mountains and rocks, and horned animals tore the earth, forming valleys. The people had come out of their Mother's womb to live with other creatures in the light of Sun, Moon and stars. Since that time they have known that they are dependent upon Sun Father, that people cannot live forever, but that their passing will result in life-sustaining moisture and opportunity of life for others.

The cycle of life brought rejuvenation of the land. Puebloan Indians of the American Southwest were carefully attuned to the migrations of the sun in the sky throughout the year and the drastic changes that took place seasonally. They also had available a landscape that offered many useful signposts. The autobiography of a Hopi man, for example, contains the following passage:

Another important business was to keep track of the time or seasons of the year by watching the points on the horizon where the sun rose and set each day. The point of sunrise on the shortest day of the year was called the sun's winter home and the point of sunrise on the longest day its summer home. Old Talasemptewa, who was almost blind, would sit out on the housetop of the special Sun Clan house and watch the sun's progress toward its summer home. He untied a knot in a string for each day. When the sun arose at certain mesa peaks, he passed the word around that it was time to plant sweet corn, ordinary corn, string beans, melons, squash, lima beans, and other seeds. On a certain date he would announce that it was too late for any more planting. The old people said that there were proper times for planting, harvesting, and hunting, for ceremonies, weddings, and many other activities. In order to know these dates it was necessary to keep close watch on the sun's movements.

These are the ingredients for the Puebloan calendar. *Everything* is related to the sun and its seasons, and the people know that Father Sun is in charge of changing climatic conditions that make life possible and survivable.

Keeping a calendar can be done in several ways, but by far the simplest, most reliable, and most accurate is by watching where the sun rises and sets on the horizon throughout the year. Ideally this requires a specific place to watch from—an "observatory," we might say today. Also required are distant contoured horizons centered to the east and west where the sun changes its rising and/or setting points between its extremes in summer and winter (the solstices). When such an observing station is not convenient, other methods can be employed. These include the construction or placement of posts or stone pillars for marking rising or setting directions of the sun. Ports in the walls of buildings, oriented toward important directions, can allow sunlight to enter, and these can be combined with markings on walls opposite the ports. In fact, nearly any door or window on the east or west sides of buildings without obstructions in those directions can be used in conjunction with wall markings to keep accurate calendars.

All of these methods have been used by contemporary and historical Pueblo people, and there is growing evidence that ancestral Puebloans used them as well. Sun Priests are still active in many of the Pueblo villages in Arizona and New Mexico. They are, in fact, among the most important and powerful people in these villages. Astronomical practices described by the first scholars to observe Pueblo cultures are clearly preserved and defined in ancestral Pueblo ruins. Places such as Chaco Canyon in New Mexico, Mesa Verde in Colorado, and Hovenweep in Utah and Colorado have features that are aligned to the solstices and

49

LEFT: The intricate designs painted on this Navajo pot depict Father Sun and Mother Earth, central characters in the creation myth of the Pueblo people.
RIGHT: Carved by a Haida artist, this massive yellow cedar statue shows Raven releasing the first people from a giant clam shell. Both creator and trickster, Raven plays a role in many stories about how the world works.

equinoxes. When these are combined with ethnographic materials that specify the roles of astronomy in the lives of Puebloan peoples, we become aware of the deep significance of the calendar as a governing agent in the lives of people living in the American Southwest. Variations are endless—as endless as variations in culture.

As an identified culture, the Pawnee Indians lived their traditional lives in the center of what is now the lower continental United States, currently southern Nebraska and northern Kansas. Of earlier periods in this culture we know very little. Their language places them in the Caddoan stock, suggesting that they had moved into the plains from the south. They lived along the Platte and Republican River systems until the pressures of European peoples spreading across the continent caused them to move to Oklahoma Territory in 1874–76. Living in earth lodges, they performed ceremonies, planted and harvested corn, beans, and squash, and hunted bison and other wild game.

There are four subdivisions, or bands, making up the Pawnee tribe, one of which is the Skidi, or "Wolf," band. The Skidi believed that their ancestors had literally come down from the stars, and they watched both Earth and sky to know when and how to do the things that brought successful living in their particular part of the world.

In the beginning Tirawahat gathered the star gods together, placed them in their proper stations, and gave them powers to create people. He directed the Great Red Morning Star and the Sun to stand in the east, and the beautiful White Evening Star and the Moon in the west. These would be the parents of the first people. He put the star-that-does-not-move, the Chief Star, in the one spot where it could always watch over the band of stars to be certain that they moved as they should. This would be the model for chiefs among people when they were put on Earth. He placed other stars and among them a pathway for the spirits of the dead to travel to the hereafter. Four stars were placed in the intercardinal directions as pillars to hold up the sky, as a model for the construction of the lodges the people would live in and to represent important plants, animals, seasons of life and of the year and other things.

Tirawahat instructed the Great Red Star to make a journey from his male, east side of the heavens to the female, west side to court the beautiful Bright White Star. The journey was difficult, filled with dangers, but the Great Star did as he was told and succeeded in mating with Bright Star. From their union came the first human, a girl, who was sent down to Earth. Meanwhile, from the union of Sun and Moon came the first male child. From these two, the human race came into existence....

When the snow began to melt, a pair of stars, the Swimming Ducks, made their appearance in the southeast just before the Sun brightened the sky. Then the Priests listened for the sound of thunder rolling from the west across the Plains. When both the calendar-setting stars and the storms were present, the bundle of the Evening Star would be brought down from the wall, opened, and the round of ceremonies would begin. In proper sequence, all the ceremonies would be performed as the people planted their crops and went on two buffalo hunts each year.

This encapsulates the world view of the Skidi Pawnee Indians, who lived in a climate where the seasons were very pronounced, and at a latitude where the intercardinal points (those associated with the solstices: northeast, southeast, southwest, northwest) were par-

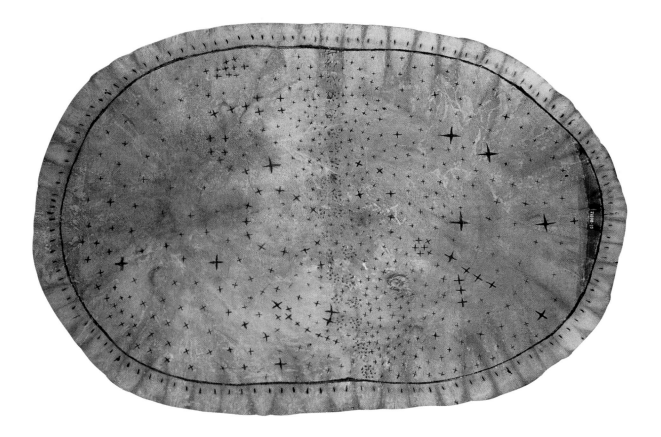

ticularly useful in seasonal reckonings. The story of how the Great Red Morning Star courted the Bright White Evening Star to bring humans to Earth seems to be based upon an awareness of the movements of planets. We know that the Skidi Evening Star is Venus, and the best information we have indicates that the Great Red Morning Star is Mars. The pattern of the Skidi story is that of the movement of the red planet Mars in its complete transit across the sky, a period of about 18 months. It begins as Mars is first visible in the eastern glow of morning light, and continues as it rises earlier and earlier and gets brighter and brighter until it is rising at sunset. Then it begins to fade as it continues to move until it reaches the western sky at eventide. When it reaches the western sky and is disappearing in the glow of sunset, it might or might not come into conjunction with brilliant Venus. When it does, it seems to reenact the mating of these two celestial parents of the Skidi people, the pair leading to the first female child on Earth. The continual cycling of the moon past the sun each month provides the observational basis for the birth of the male human of Skidi cosmology.

Each February the Skidi new year began when the Swimming Ducks, the two stars marking the stinger of the southern constellation Scorpius, made their appearance in the sky before morning twilight. At this time, the Council of Chiefs, Corona Borealis, was high over-

The Skidi, a band of the Pawnee tribe of the Great Plains, made this hide star chart.
They believed that their ancestors descended from the sky. Venus and Mars gave
birth to the first female; the sun and moon produced the first male.

head, ratifying the beginning of the ceremonial year. On these February evenings, the Pleiades also stood near the zenith as darkness descended, verifying for the Skidi the importance of being united throughout the coming year. This was the time when the Skidi priests listened for the sound of thunder as spring storms began moving across the prairie. This combination of astronomy and meteorology, along with observations of plants and animals, is a superlative example of a calendar system designed for successful living in a widely changing environment.

The Olmec, a well-organized culture that flourished in southeast Mexico and other lands along the Caribbean Sea from 1200 to 400 B.C., developed a complex method of monitoring time that was based on two calendars. From observations of the sun and stars they knew that the agricultural year consisted of 365 days. They used this information to constantly recalibrate their ritualistic calendar of some 260 days. This duality was inherited by the Maya, who rose to power between A.D. 250 and 950. This culture extended from Copàn (Honduras) northward to Tikal and Uaxactun (Guatemala), westward to Palenque (Campeche), farther northward into Yucatán to such places as Uxmal and Chichen Itza, and still westward to Teotihuacán (near Mexico City). Many of these places show evidence of extensive astronomical knowl-

The first predawn appearance of the Swimming Ducks (left, near the horizon), the
pair of stars that form the stinger of the constellation Scorpius, signified the
beginning of the Skidi new year in February.

edge among the people who built some of the greatest and most complex ceremonial sites ever recorded. Impressive cities with delicate sculptures and monumental architectural structures, including pyramids topped by temples decorated with ornate sculpture, reveal a people obsessed with numbers and calendars, and capable of comprehending and calculating time backward and forward millions of years.

The rich Aztec mythology, recorded by Spanish chroniclers, fills out the story. When the Spanish arrived they destroyed much of the Aztec civilization and uncovered the remnants of the Maya. Diego de Landa, first archbishop of Yucatan, proudly described how they destroyed all things relating to the Mesoamerican religious beliefs, which the Spanish found so demonic. Concerning the many glyphic books they found that recorded the history and beliefs of the people, he wrote:

> We found a large number of books in these characters [hieroglyphs] and, as they contained nothing in which there were not to be seen superstition and lies of the devil, we burned them all, which they regretted to an amazing degree, and which caused them much affliction.

Only fragments of Maya books, printed on bark paper, survived and found their way into western hands. These are filled with Maya astronomy—especially lunar and solar almanacs and a detailed ephemeris of the planet Venus. From these few books and from the writings engraved upon buildings and stone monuments, scholars have been able to learn a great deal about Mesoamerican cultures, including their obsession with cycles of time. All these cycles, involving agriculture, ceremony, and Venus observation, were bound together in the Maya way of measuring time and thinking about their own origin, history, and destiny. The Maya knew, for instance, that every eight years, or 2,920 days (8 x 365), Venus phenomena would reoccur because eight years is equal to five complete Venus cycles (5 x 584 = 2,920 days). This eight-year period of time was of great importance to Mesoamerican cultures. The reappearance of Venus as Morning Star after its brief disappearance from the evening sky was particularly important for rituals involving warfare and human sacrifice. Knowing the observational cycle of Venus and being able to accurately predict this cycle was of the utmost importance. Every 584 days Venus would make its rapid transit from evening to morning sky, and 2,920 days after one of these Morning Star appearances Venus would dash into the morning sky to stand among the same constellations where it had been eight years previously.

An even longer cycle of time, which a typical individual might experience only once in

High overhead in the late-winter sky, the Council of Chiefs—visible as the close-knit star group, the Northern Crown—presided over the start of the Skidi ceremonial year.

53

If you look just north of east and about halfway up in the sky in early October about an hour past sunset, you will find a lovely little cluster of seven blue stars that, if your eyesight is very good, looks like a tiny fuzzy dipper. Many people think so when they first see it. Some even think it is one of the Dipper constellations. Go to a place far from city lights on a night when there is no moon in the sky, and the little fuzzy collection stands out dramatically like a shining jewel. There is nothing else quite like it in the night sky. Through a set of binoculars it is spectacular, with dozens of blue stars becoming visible, and through a telescope, it is breathtaking, with too many stars to count.

This is the Pleiades, to which more stories and intriguing tales are linked than to any other collection of stars. They have been visualized as young women, old women, boys, birds, chickens, camels, dogs, goats, raccoons, fish, rattles of a snake, seeds, a bunch of balls, crystals, eyes, and even piñon nuts. A replica of an ancient Hawaiian sailing canoe, the "Makali'i," was named for them, and so was an automobile, the Subaru, which uses the star cluster as its logo.

Long before our founding fathers penned the words that defined the United States of America, Pawnee Indians looked to these tightly bunched stars to formulate an earlier concept of unity with the land. Navajos carved them on rocks, put them on sand paintings and gourd rattles, and performed most ceremonies during the part of the year when they are present in the evening sky. On the Colorado Plateau seeds must be sown between the time when these stars vanish in evening twilight in April and when they reappear before morning light in June. In many ways, the Pleiades has guided people throughout the world, and they have provided clues for modern astronomers about how stars form, why they shine, how long they live, and when they die.

The Pleiades (M45, or the 45th object in the Messier Catalogue) is a physical cluster of stars that were born together out of a huge cloud of gas and dust some 100 million years ago. Lying at an average distance from Earth of 380 light years, some of the brightest members of the group (the seven "sisters") have been found to be rapidly rotating hot young stars. Some, like Pleione, are rotating so fast, some 100 times faster than the sun, that they probably look more like doorknobs (prolate spheroids) than spheres and from time to time have ejected vast amounts of material into space. Lying near the ecliptic, the Pleiades is periodically occulted by the moon, which is a beautiful sight in a good pair of binoculars or small telescope.

—*Von Del Chamberlain*

A brilliant, compact cluster of blue stars, the Pleiades occupies an important place in the sky lore of cultures around the world. It figures in many Native American legends and beliefs, and is pictured in rock carvings and on ceremonial objects.

COSMOLOGIES OF THE AMERICAS

THE CLASSICAL UNIVERSE

COSMOLOGIES OF THE AMERICAS

a lifetime, marked a complete cultural life cycle. The Maya believed that the world had been destroyed and recreated several times in the past, and that it might again be destroyed. A particular 52-year cycle was closely watched for this reason. Why 52 years? Because that is the period during which the combined 365-day and 260-day calendar had to start over. The number of days in 52 years is 18,980. This is equal to the number of days in 73 ritual years of 260 days each. Thus, combining the agricultural and ritual calendars, every day of the 52-year cycle had a unique designation, but after 52 years the combinations were exhausted and the designation of the days had to begin anew. As the end of a 52-year cycle approached, people cast away old things, cleaned their houses, and put out their fires. The astronomer priests climbed the "Hill of the Star" and watched the Pleiades climb up the sky toward the zenith at midnight. They believed the world would end if the Pleiades stopped, but once it was clear that this tiny and distinctly prominent grouping of stars reached the western half of the sky, they knew that the world was safe for at least another 52 years. New fires would then be kindled in the central temples and their embers spread throughout the land to all the villages to start new fires amid celebration of a new beginning. These Mesoamericans also knew that two 52-year cycles were commensurate with a complete Venus cycle: 104 years equals 37,960 days, which is the number of days in 146 ritual years (2 x 73 x 260), the same number of years necessary for 65 Venus cycles (65 x 584). This is one of the longest cycles known to date in the Americas, spanning lifetimes and requiring a structured, hierarchical culture to support it.

Encountering the cosmologies of Native American cultures has broadened and deepened my appreciation of the astronomical universe as a reflection of the human condition. My formal training in astronomy opened up the physical universe to me, but my first contact with the cosmologies of native cultures, especially those of the Pueblo and Skidi, opened up an equally rich and satisfying universe, an inclusive universe encompassing all mankind. I have been able to study only a few of the hundreds of identifiable cultures that have lived on the American continent, and what impresses me is that they all used the sky for orientation and guidance. The cosmologies they created are intriguing interpretations of the world they perceived and reflect the differences in their environments. This realization has helped me appreciate the origins of our own ways of looking at the universe in Western culture. Western science always tests its own creation myths, modifying them to suit new observations. Sometimes the modifications are incremental, sometimes revolutionary. Concepts such as the primordial atom and the big bang, dark matter, dark energy, and superstrings all are creations designed to make the puzzle pieces fit together better, to give us a better understanding of our place in the universe. All cosmologies, traditional and physical, are human inventions drawn from minds in touch with the universe, and they give us insights to help us appreciate ideas about human origins and histories and our relationships to the cosmos.

PRECEDING PAGES: Throngs of people celebrate the spring equinox by climbing the Pyramid of the Sun at Teotihuacán, Mexico, the site of an ancient Toltec sacred city. Mesoamerican cultures developed an impressive knowledge of astronomy.

Mars, Jupiter, and Venus (top to bottom) are seen in a rare planetary alignment above a Maya temple at Tikal in Guatamala. Planetary cycles played an important role in many aspects of Maya life, from agriculture to the timing of religious rituals.

AFRICA'S MORAL UNIVERSE

Christine Mullen Kreamer

Turn out the lights and watch the real ones in heaven—
those our ancestors' imaginative minds used
to mold a wonderful poetic imagery about themselves
and their relation to the universe.

—Anthony Aveni

Much of my world has been Africa. I have long been fascinated by African thought systems and how they operate within the complex, changing contexts of contemporary African life. Since my student years, I have been interested in the visible expression of African cosmology and how objects and actions symbolize and, in a sense, "perform" a society's moral universe. Africans, I have discovered, creatively experience the universe in their everyday lives.

My professional journey of discovery began in the early 1980s

A total solar eclipse darkens the sky above a thorn acacia tree near Chisamba, Zambia, on June 21, 2001. While such an event is rare, the sun, moon, planets, and stars play a central role in African life. The diverse cosmologies of Africa find expression in architecture, sculpture, performance arts, and everyday rituals.

when I embarked on field research with the Moba peoples of northern Togo in West Africa. As a fresh-faced Fulbright scholar newly arrived in northern Togo, I was delighted to be invited to attend a ceremony of initiation into the secret society of *Kondi*. Some four hours' walk later, I found myself in the village of Tonte and, for a brief time, the center of attention as the initiate and her attendants gathered around me, greeting me in an astonishingly deep, gutteral chant that made the hairs on the back of my neck bristle. Then the initiate moved on. With her body bent, a wooden hunting stick in one hand, and her face covered with twisted leather strands she held in the other, she scampered through the parched landscape of the late dry season in a manner reminiscent of the bounding gait of an antelope—one of the important animals in Moba creation myths about the founding of community and the ownership of specialized knowledge.

As the rest of the ceremony unfolded, I witnessed rites of passage depicting a symbolic death and then rebirth as she began her reintroduction to the community as an adult. This is a common theme in African coming-of-age rituals that emphasize ideal moral qualities and social values, some of which are tied to concepts related to the sun, moon, and stars.

Within several days, the *Kondi* had christened me Kondjiitmang—the red Kondjiit (an obvious reference to my sunburned face), a name that has remained with me among my Moba family and friends ever since. It was the beginning of a lifelong inquiry into the many ways that African sculpture, architecture, and performance arts reflect the beliefs, values, and cosmologies of diverse African cultures.

In Africa, as throughout the world, beliefs about the cosmos are linked to a society's oral traditions, myths, and symbolic systems. Cosmology may include knowledge of the structure of the physical universe, but it goes far beyond that to include the beliefs and values that sustain the community, its notions of creation, the structure of time, and the realm of the ancestors; and its organization of everyday and ritual space. In Africa, cosmologies are connected with indigenous knowledge and with concepts about the ideal person, primordial ancestors, leadership paths, relations between the living and the dead and between humans and the natural world, and home and community as microcosm for a society's worldview.

The stars and planets have been a source of inspiration in the creation of African arts for thousands of years. At the time of the pharaohs in ancient Egypt major deities were associated with the sun, the moon, and the Earth. Today, throughout Africa, time-reckoning, the seasons for planting and harvest, and the timing of various rituals are still based on the positions of the stars and planets. The Borana peoples of Ethiopia follow a lunar calendar based on the appearance of the crescent moon, and lunar calendars in many other African societies are used to regulate agricultural work. Sirius, one of the brightest stars in the Southern Hemisphere, is noted in the astronomical observations of a number of African societies and may well serve as a celestial "anchor" in calculating time and space, but more research into tradition-based systems of astronomy in Africa is needed. Nonetheless, it is clear that observations of the heavens are part of the knowledge that influences the construction of social

institutions, artistic expression, and ritual practice in African cultures, as throughout the world.

In the creation myths of many African societies, the first beings descended from the sky, the home of powerful spirits and a creator god. These original ancestors, who were often credited with extraordinary powers and the founding of communities, served as intermediaries between the realms of earth and sky. Among the Dogon of Mali the original eight founding ancestors, or *nommo,* are masters of water and rain who came to earth from the sky and stars. They are often carved with their arms raised in a gesture connecting their celestial home to the earthly realm to which they descended. Their roles as mediators between earth and sky—between humankind and Amma, the Creator—are also illustrated in carvings of stools atop which human figures perch but do not sit. According to Dogon belief, such stools represent the sky supported by the mythic ancestors. As such, they are not suitable as functional furnishings for human beings.

A similar connection between earth and sky can be found in the framework of the Dogon *kanaga* masks used in funeral dances to lead the soul of the deceased to the afterlife. The upper crossbar of the *kanaga* mask is said to represent the sky and the lower bar the earth, emphasizing the notions of duality that are part of the Dogon conception of the world. Dogon masks and figure sculptures are carved by blacksmiths who, as metalsmiths, are also masters of earth, air and fire.

Dogon *kanaga* dancers in Mali wear tall masks during funeral dances to guide the soul of the deceased into the afterlife. The upper crossbar of the mask is said to represent the sky; the lower crossbar, the earth.

He carries a heavy stone upon his head without a cushion.
Shango splits the wall with his falling thunderbolt.
   —Yoruba praise poem for Shango, the thundergod

Among the Yoruba of Nigeria, the sky is the realm of Shango, the god of thunder and the legendary sacred king of the ancient city of Oyo-Ile. He was known as a powerful warrior and magician, a hot-tempered deity linked with fire and water whose earthly presence is revealed in the "thunderstones," or stone axes, that he hurls to Earth during thunderstorms. Despite his mercurial temperament, Shango is a popular deity associated with both creation and destruction. In ceremonies that honor Shango, priests and priestesses dance with wooden staffs that depict his symbol of the double-bladed axe, which is sometimes rendered as a headdress with two projections. Shango staffs typically include a kneeling female figure, her gesture suggesting an offering. This female devotee serves as a "cool" counterbalance to the "hot" virility of this powerful male deity, a common pattern found in much of African art, which illustrates the complexities of the human experience in terms of duality or opposites.

Oddly, the sun is rarely depicted in tradition-based African arts, but it is not unimportant in African beliefs. The sun's life-giving properties are well appreciated—to such an extent that, in many African languages, the words for the sun and the supreme being are the same or similarly derived. In the language of the Moba of northern Togo in West Africa, for example, *yendu* means both the creator god and the sun.

Among the Kongo of the Democratic Republic of the Congo, the sun's journey across the sky is likened to the passage of the human soul through birth, maturity, death, and rebirth. The rising and setting of the sun also symbolize the relationships between humans and the divine and between the living and the dead. Kongo artisans express this in the form of a symbolic chart of the soul, or cosmogram, called a *dikenga,* which is often depicted as a cross inside a circle, or as a simple cross or diamond. These designs represent the four moments of the sun as well as the continuity of human life and the spiritual crossroads where human beings and the spirit world meet and communicate. When used in Kongo rituals, this cosmogram suggests insight into both worlds. It also demonstrates how natural phenomena, such as the movement of the sun, become part of the philosophical underpinnings of human experience.

The first work of mankind is farming.
   —Bamana proverb, Mali

Farming continues to be a predominant occupation in Africa. The sun's importance to a successful harvest is acknowledged in many ways, most dramatically by antelope crest pieces carved by the Bamana of Mali. These depict *chi wara*, the mythic ancestor—half human, half roan antelope—who taught the Bamana how to farm. During ritual dances honoring *chi wara* for bringing this knowledge, Bamana men wear these crests attached to basketry caps. The crests are carved in pairs—male and female—to underscore the essential belief that men and women must work together to secure a bountiful harvest. The openwork mane of the male antelope is said to represent the path the sun takes as it moves across the sky each day.

Its horns are compared to the stalks of millet (a staple food for the Bamana) that grow straight and tall under the warmth of the sun.

The moon also plays a role in the success of the harvest. In many African societies, it is a metaphor for human fertility and, by extension, agricultural fertility. Lunar cycles form the basis for agricultural calendars throughout the world. In Africa, masked performances may also mark the phases of the moon. A particularly dramatic mask attributed to the Sungu-Tetela peoples of the Democratic Republic of the Congo was reportedly used for the dance of the new moon, as well as for funerals and other occasions. Originally, the crest of the mask held vulture and guinea fowl feathers, and the costume included a raffia skirt worn over a leopard or spotted cat skin. When worn together, the crest and the skirt symbolized notions of power that combined both earthly and celestial forces.

In a variety of African creation myths, both sun and moon are gifts from the creator that bring the fullness of life, hope, and promise to the world. As reported by the anthropologist Allen F. Roberts, in the creation myth of the Tabwa of the Democratic Republic of the Congo, the Earth was originally a cold, dark place without vegetation. The creator god, Leza, sent the first man and the first woman on a long journey to populate this barren land, giving the woman a basket of gifts to take with her. After their long journey, the couple stopped to rest. The man found some wood and built a fire, and the woman opened her basket. When she did so, the sun emerged and rose to the heavens to warm the Earth. When evening came and the sun had set, the woman again opened her basket, releasing the moon and stars to the sky.

65

The phases of the moon can also denote philosophical opposites, such as understanding and confusion, fortune and misfortune, or goodness and evil. This is the case among the Tabwa, who see the rising of the new moon as a sign of hope and rebirth, and the disappearance of the moon as a metaphor for envy, greed, and the desire to do harm. Roberts reports that in Tabwa philosophy, the phases of the moon reflect the duality and complexities of the human condition. Figure carvings of Tabwa ancestors bear scarification patterns that include parallel isosceles triangles called *balamwezi*, literally "the rising of the new moon."

Sungu-Tetela dancers in the Congo wore this helmet mask during funerals, marriages, and performances to mark the occurrence of the new moon.

Carved wooden panels depicting men, women, and animals were created in the early 20th century by the Congo's Nkanu peoples as part of their initiation rites. Background designs sometimes depicted the sun, but more often were ornamented with the moon and stars, which served a symbols for fertility.

The sun, moon, and stars also find their way into the arts of Ghana's Akan peoples. The stamped designs of the *adinkra* cloth worn at funerals include a number of celestial bodies, and these images communicate the wisdom of appropriate behavior within the community.

Sirius, the brightest star in the heavens, features prominently in the astronomical lore of many African cultures. One of the more controversial interpretations of African knowledge of the Sirius system concerns accounts of Dogon lore published in the mid-20th century by French anthropologists Marcel Griaule and Germaine Dieterlen. The anthropologists asserted Dogon knowledge of Sirius B (*po tolo*, *tolo* meaning *star*), a white dwarf that is a companion star to Sirius (*sigi tolo*), but is invisible to the unaided eye.

Their research among the Dogon documented the important celebration called Sigui, held every 60 years or so and said to occur upon the completion of Sirius B's journey around Sirius A and with the synchronized orbits of Jupiter and Saturn. The Sigui ceremony includes spectacular masquerade performances and recalls Dogon creation myths. Dogon tradition says that po tolo, or Sirius B, is "the egg of the world" and the beginning of time. Its orbit around Sirius A is counted twice by the Dogon, fitting into the concept of duality prevalent among the Dogon and other peoples in the region; *po* is the Dogon word for the tiny grain of millet, a staple cereal crop grown by the Dogon.

Is it possible for the Dogon to know of po tolo without the aid of some form of telescope? Was knowledge of Sirius B introduced from the outside—perhaps inadvertently by missionaries or even by Griaule himself who studied astronomy and may well have asked a "leading question" that fit with Dogon experience—and, thus, incorporated into Dogon cosmological lore? Or, as some scholars have suggested, was knowledge of

Sirius transmitted to the Dogon through the ancient Egyptians, and knowledge of Sirius B conveyed by extraterrestrials from the Sirius system itself, as suggested in a book by Robert Temple called *The Sirius Mystery?* In the presence of such diverse interpretations and in the absence of conclusive evidence, the question of Dogon knowledge of Sirius B remains a mystery and will, no doubt, continue to generate debates over astronomical knowledge and the power of suggestion. The answer might well lie somewhere in the middle—a mix of indigenous astronomical observations and culture contact.

To astronomers, Sirius B was as much a mystery as was Dogon knowledge of its existence. It was discovered by chance in 1862 when the American telescope maker Alvan Clark turned his latest creation, a magnificent 26-inch refractor, toward Sirius just to test the quality of the lens. Because Sirius formed a nice visual double star system, within a few decades the combined mass of the system was measured and the two stars were found to be similar in mass (a factor of 2), even though B's visual brightness was only 1/10,000 that of A, its bright companion. This puzzled astronomers for decades. The laws of physics demanded that it be extremely small and dense to appear so dim. This was explained in the 1920s and 1930s when theoretical astrophysicists such as S. Chandrasekhar were able to show mathematically that stars such as Sirius B, now called white dwarfs, are in a physical state far different from anything we are familiar with on Earth.

—*Christine Kreamer*

This x-ray image of the Sirius star system, by the Chandra x-ray satellite, reveals Sirius B as the brighter central star, with Sirius A above it. Because Sirius B's atmosphere radiates at a temperature of 25,000 kelvins, it emits x-ray radiation. Sirius A is vastly brighter than B in the visual region of the spectrum. It therefore appears here to be fainter than B because it radiates weakly in the x-ray, though it is brighter in a neighboring region, the extreme ultraviolet, which probably leaked through Chandra's filters to produce this dramatic image.

The star is seen as a child of the sky. As R. S. Rattray noted in *Art and Religion in Ashanti* (1927), stars communicate the importance of faith and reliance on others: "Like the star, the child of the Supreme Being, I rest with God and do not depend upon myself [alone]." Another popular *adinkra* pattern is the altar to the Sky God, Nyame. The crescent moon appearing in this design is a feminine symbol associated with faithfulness; when combined with a star, the *adinkra* design conveys the interdependence of men and women in marriage. The swastika, an ancient sun symbol for many cultures throughout the world and seen in many variations, is yet another *adinkra* symbol signifying the importance of serving others, for it is identified as the hairstyle worn by the Queen Mother's attendants.

The Earth Goddess fashions the human body just as a potter fashions her pot.
—Jukun proverb, Nigeria

African societies associate the earth with physical sustenance, the continuation of the human race, and the fertility of the land. This has its roots in the economic importance of agriculture in much of rural Africa. Because a successful yield depends upon a combination of skill and environmental knowledge, these are codified by rites and ceremonies dedicated to the earth, which is often personified as a feminine deity or spirit. The sacred qualities of Mother Earth and her life-giving capacities mean that those who work with earth—such as potters and farmers—embark each day on a creative process that puts them in touch with the sacred. In the case of potters, manipulating earthen materials and fashioning pots is likened to birth and the fashioning of human beings.

African notions about the earth demonstrate a deep respect for the environment and are part of long-established indigenous conservation practices dedicated to protecting and preserving Mother Earth. Common throughout much of West Africa's savanna is the institution of Custodian of the Earth, an inherited religious and political office. Earth Custodians are given the responsibility of maintaining the purity and potency of the earth so that humankind can continue to prosper. Among the Moba of northern Togo, earth priests called *tingdana* perform regular offerings at small earthen and thatched roof shrines A symbol of their power is an earth-packed animal horn containing the spiritual potency associated with the earth and with revered ancestors. Custodians of the earth also exercise control over the agricultural calendar and ensure that some sections of the Moba landscape remain uncultivated, designated as shrines to the earth. Increasingly threatened due to population pressures on the land, these sacred spaces forge a spiritual connection between humans and the divine, and serve as local environmental reserves that are a haven for precious plant and animal life.

In Nigeria, for the Yoruba peoples who belong to the Ogboni association, the earth, Ilè, is conceived as a mother goddess and the source of moral law. Her powers are conceptualized in cast-metal figures or heads, one male and the other female, joined by a chain and usually ending in a metal spike. They are insignia of rank owned by those who have been inducted into the Ogboni society and members often wear them around their necks or plant them in the earth to emphasize their link to the earth goddess. An excerpt from a Yoruba praise poem to Ilè the earth goddess illustrates the deep respect African societies have for the earth and her moral laws:

68

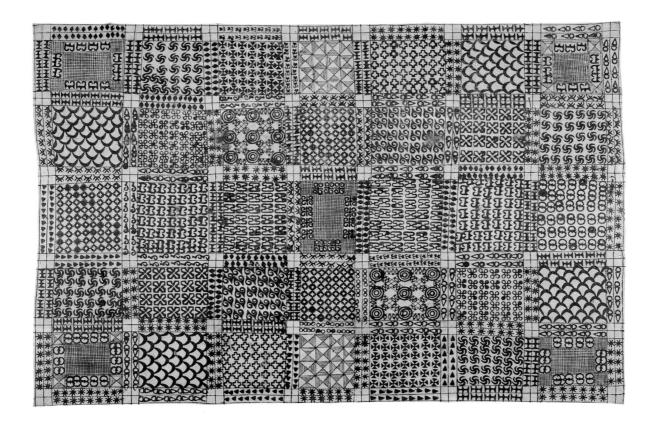

*. . . Earth is the mother of the "one
who wakes up to meet honor,"
otherwise known as Edan
May we not step on you with the
wrong foot
May we step on you for a long time
For a long time will the feet walk
the land
May we not step on you, Earth,
Where it will hurt you.*
—Yoruba praise poem for Ilè

Myths about Mother Earth—the personified giver of life—place her within a community that parallels human society. Among the Igbo of Nigeria, Ala is the earth mother and Amadioha the sky father who, together with their spirit children, create the world and the Igbo concept of community. One of the more spectacular ways that Ala and Amadioha are represented is in Igbo *mbari* houses—polychromed earthen constructions—that include wall paintings, relief carvings, and three-dimensional sculptures featuring deities and scenes from everyday life. The power of *mbari* houses is enhanced through the use of earth taken from termite mounds, from which human life is believed to be reincarnated. In many

Rich in symbolism, the designs found on *adinkra* cloth of the Asante peoples of Ghana often include references to celestial bodies. Traditionally worn during periods of mourning, the cloths convey a community's shared values and beliefs.

AFRICA'S MORAL UNIVERSE

parts of West Africa, the earth of termite mounds is considered particularly potent, and therefore an appropriate ingredient for shrines and offerings, because it comes from the abodes of creatures who link earth and sky, and figure prominently, as earthworks "architects," in the myths of certain African cultures. *Mbari* houses, made most often as offerings to Ala and in response to community problems, have been likened to spiritual places where the Igbo community of humans and divine beings coexist. For the Igbo, the act of creating *mbari* houses is a community sacrifice to renew the local world and prepare for the future. Ala is also the patron of Igbo potters, who use earth in forming pots for everyday and ritual purposes.

The mystical powers of the earth are brought to bear in the construction of Kongo power figures, or *nkisi* (Democratic Republic of the Congo), which use earth to enhance the potency and efficacy of the carving. Earth from graves, gullies, or streambeds, which is thought to be infused with the spirit of the dead, is often inserted into cavities in the figure or packed around significant areas of its body. The abdomen, which is associated with life and the soul, and the head, as the site of communication with the spirits, are particularly appropriate for receiving powerful medicines. Combined with other natural and manufactured ingredients, and enhanced with the invocations of the ritual specialist, or *nganga*, the carvings are mixed media *tours de force* that function as containers for spirits from the other world.

Even the organization of living space reveals a relationship with the cosmos. According to art historian Suzanne Blier, the two-story earthen houses of the Batammaliba of northeastern Togo and northwestern Benin are a cosmological model. Their upper regions are associated with the heavens; the section below, supported on cross beams, is linked with the earth and human activity; and the ground floor is for animals, ancestor shrines, and beings of the underworld. Batammaliba houses are oriented toward the western winter solstice, the most critical point in the sun's path in the course of the year, marking when the days must begin to lengthen again. To ensure that this happens, the houses become sanctuaries dedicated to Kuiye, the solar deity.

Among the Moba of far northern Togo, the east is associated with men, the work of farming and hunting, and the rising sun. The west relates to women, the work of carrying water and cooking, and the setting sun. Moba earthen compounds are constructed so that their openings face the setting sun, thus welcoming the prosperity and good fortune that are associated with Yendu, the solar creator god. To the Moba, the home is also a metaphor for the human body and, by extension, for the social body of the community and its relationship to the moral universe created by Yendu. This is particularly the case in Moba funeral rites, where the house serves as a surrogate for the deceased, whose clothing is displayed on the roof. During a man's funeral, he is "attacked" in stylized displays of aggression through attacks on his home. Miniature arrows are shot over the compound wall to emphasize the tensions surrounding death and to encourage the deceased to leave home and join the spirit world. The house is also the theatrical backdrop to the music, dance, and ceremonies that celebrate the life and mourn the death of a respected member of the community and prepare the way for the deceased to become a revered ancestor.

The cosmos and notions about the origins of the universe play a central role in the way

Africans define themselves, their moral values, and their place within the world. Far from abstract, removed concepts, African notions of the universe are intensely personal and place human beings in relationships with the sun, moon, stars, and earth. Standing at the core of creation myths and at the foundation of moral values, celestial bodies are often accorded sacred capacities and are part of the "cosmological map" that allows humans to chart their course through life.

African arts—verbal, visual, performing—are central to articulating this important relationship between humankind and the universe. While formal properties visually identify and symbolize prominent features associated with the identities and qualities of celestial bodies, the verbal arts of myth, folklore, proverbs, songs, and ritual invocations bring them to life within a diversity of contexts—such as shrines, offerings, ceremonies, funeral celebrations, and agricultural rites. African works of art can be said to "perform" the moral universe, reinforcing through poetic imagery key concepts about human morality as it is linked to the ideal and to the divine.

71

One memory from my time in Africa stands out as personal proof that the intimacy of the sacred cosmological world can exist side by side with modernity in the contemporary mind. For the Moba of northern Togo, sighting an especially bright shooting star is a sign that a great chief has died. I saw only one such shooting star during my years in northern Togo and, sure enough, a few months later, commemorative funeral rites for a respected Moba chief from a neighboring village were held at the auspicious time of the full moon. I had the honor of being asked to attend. After the ceremonies had ended, I set out on foot to

Carved Kongo power figures, called *nkisi*, serve as containers for spirits from the other world.

make the three-hour trek back to my home village. It was too late to turn back when I realized that the light of the moon had significantly diminished. Looking over my shoulder, I discovered that we were experiencing a full lunar eclipse. By the time I reached my home in the village of Nano, my neighbors had already realized what was happening. They were all outside banging drums and pans, and the children were singing, over and over, "The sun ate the moon! The sun ate the moon!"

The festive display had a serious purpose: It was designed to entice the sun to give back the moon that it was devouring, and it continued until the moon was restored in its entirety. Only then was order achieved and the Moba world restored to its proper balance.

STAR MAPS:
A CONFLUENCE OF
ART AND ASTRONOMY

Deborah Jean Warner

When you look through a telescope, what do you see? And how do you communicate that information to someone else? This is an especially important problem for scientists, given the impact their findings often have on all of us. I see star maps and telescopes as equally important scientific instruments. Both require certain skills to operate, and they both require a common language to make the information they provide useful. For the scientists who designed these star maps—first created before telescopes were turned to the heavens and produced through the 19th century—star maps were a way of communicating what they knew about the universe to others. But these maps also serve a purpose that scientists never planned or envisioned. Centuries after they were made, they exist as a reflection of how

In this 1504 woodcut by Albrecht Dürer, the Arabian astronomer Messahalah views celestial objects through an overhead dome to provide a frame of reference, then plots their positions on a globe. For centuries, celestial maps have provided a way for humans to impose order on the universe.

STAR MAPS: A CONFLUENCE OF ART AND ASTRONOMY

the individuals who made them viewed life, the world, and society, as well as the universe as it was known to them.

I have always been fascinated by the connections between art and science, so I was thrilled to discover, soon after coming to the Smithsonian as an assistant curator in the 1960s, a small stash of early modern star maps in what is now the National Museum of American History. At the time, little was known about the origins and meanings of these beautiful and intriguing documents. This was quite a mystery, and I couldn't wait to start looking for the missing pieces of the puzzle. As it happened, this was a good time to tackle this project. Historians of science, who had long valued the growth of ideas above all else, were beginning to investigate the history of scientific instruments and experiments, and art historians too were beginning to broaden their scope, looking beyond the traditional fine and decorative arts. My research eventually led to the publication of a book, and to a deeper appreciation of the cultural roots of celestial mapmaking.

MAPPING THE HEAVENS

The Western tradition of celestial cartography has two roots. One lies in the stories that the ancients told about the constellations. The best known constellation stories are those presented in the *Phaenomena*, a Greek poem written by Aratus of Soli in the third century B.C., and in the *Poeticon Astronomicon*, a Latin poem written by Caius Julius Hyginus about 200 years earlier. These stories were also presented in graphic form, as can be seen on the so-called Farnese Atlas, a Roman copy of a Greek statue of Atlas holding a celestial globe. This statue, unearthed in Italy in the early 16th century and acquired by the Roman art patron, scholar, and Cardinal, Alessandro Farnese, is the only surviving celestial globe from classical antiquity.

The other root of Western celestial cartography lies in the early star catalogs, particularly the one Claudius Ptolemy of Alexandria compiled in the second century and included in his *Almagest*, written around A.D. 150. Ptolemy cataloged 1,025 stars that were visible from the Mediterranean and bright enough to be distinguished by the naked eye. He grouped these stars into 48 constellations and indicated the position of each star in these constellations (by its celestial latitude and longitude), as well as its magnitude (apparent brightness), and its place within its constellation figure. This Ptolemaic catalog formed the basis for stellar astronomy for the next 1,400 years. It wasn't until the late 16th century that any astronomer recorded both coordinates of the Ptolemaic stars, or systematically charted any non-Ptolemaic ones.

Several schools of Islamic scholars continued the Ptolemaic astronomical tradition during the medieval period, and many of these scholars looked to artists and artisans to produce diagrams of the individual constellations and globes of the entire celestial sphere. Some 126 of these globes are known today. One of the largest and most wonderful was given to the Smithsonian Institution in 1974. A sophisticated work of science, it is also a beautiful work of art. Analysis of its features indicates that it was made in western India sometime between

1625 and 1650. This Islamic globe reflects a complex cultural tradition. While the 48 constellation figures are Ptolemaic and the stars are positioned as described in Ptolemy's *Almagest*, the text is in Arabic, the human figures wear Mughal dress (from the period of the Mongol empire in India), and the figurative style resembles that found in many manuscripts of the *Book of the Constellations of the Stars*, written by Abd al-Rahman al-Sufi, who worked in the Persian city of Isfahan in the tenth century A.D.

On what remains of the globe, 1,016 stars are visible. An additional three stars were probably depicted in the small area of the constellation Serpens, which is now missing. Because the stars are correctly positioned with respect to latitude and longitude (as they would have been in the 17th century), we can presume that the globemaker relied on an up-to-date star catalog that took into account precession, the wobbling of the Earth's polar axis caused by the gravitational pull of the sun and moon on its bulging Equator. Because the globe is so correct, its few errors stand out. There is, for instance, an extra star in four constellations, and one missing in four others.

The globe is also a marvel of craftsmanship. It is hollow, made of brass with the stars inset in silver, and, unlike most other metal globes, it is seamless, so we know that it was cast in one piece. Modern metalworkers have marveled at the technical skill of its maker. Measuring about eight and a half inches in diameter, it is the sixth largest of the 126 known Islamic celestial globes. It is prominently displayed at the entrance to the *Explore the Universe* exhibit at the National Air and Space Museum.

Comparable to the Islamic celestial globe in beauty and accuracy are two maps of large portions of the celestial sphere, dated 1515 and printed from woodcuts. Called planispheres, these maps

Atlas holds a cosmic sphere depicting stories the ancients told about the constellations. Known as the Farnese Atlas, this Roman copy of a Greek statue features the only surviving celestial globe from classical antiquity.

have the distinction of being the first of their kind. Three Nuremberg mathematicians collaborated on this project. Johann Stabius drew the coordinates, Conrad Heinfogel positioned the stars, and Albrecht Dürer drew the constellation figures and cut the wood blocks. Because Dürer was an artist of great influence in the Northern Renaissance and the author of treatises on a number of subjects, including perspective and the ideal in human proportions, the constellation figures are fully formed, and their classical attributes are correctly drawn.

The Nuremberg planispheres are clearly Ptolemaic: The constellations they depict are

This Islamic celestial globe, crafted in the early 17th century, is a spectacular variant of the astrolabe. It served both as a map of the heavens, as if viewed from outside the starry sphere, and as a precision tool for making astronomical calculations. Engraved on its surface are various coordinate lines, constellation figures, and Arabic inscriptions. The stars are made of embedded bits of silver and the figures are reminiscent of typical Mughal court dress. Hollow and cast in one seamless piece by the lost-wax process in Lahore (in what is now Pakistan) in the Mughal dynasty, it is the only globe of its type in the collection of the National Museum of American History.

The globe was originally set in a cradle of rings, which depicted the horizon and other astronomical circles and allowed the sphere to be useful as a calculating device. It was probably used very much like a modern educational or recreational globe, determining the rise and set time of celestial objects as they would be viewed from different parts of the Earth.

The example in the Smithsonian collection is considered to be one of the finest in existence, among some 126 known globes of this period. It is the sixth largest, with a diameter of 21.6 centimeters, and displays some 48 engraved constellations and about 1,000 stars. The globe is of sufficient historical and technical interest to have been studied in the Smithsonian's Conservation and Analytical Laboratory. Conservators there were able to determine how the globe had been constructed, finding some 74 tiny plugs in its surface that were essential in the lost-wax process of its manufacture. Radiography also demonstrated that the globe was not produced by shaping an originally flat metal disk, and that the composition was dominantly copper and lead, with traces of zinc, tin, and other metals. The globe was smoothed on a lathe before the stars were set in and the coordinate system and constellation figures engraved.

Scholars have also analyzed the engravings on the globe to determine as much about its origins as possible, and about the culture that created it. Even the hairstyles were studied, to the extent that they seemed to predate Islamic fashions; they bore similarities to the manes of the animals depicted on the globe, such as Leo, Pegasus, and Ursa Major. These in-depth analyses have made objects such as this one windows to the traditions of instrument design and construction in the Islamic world and its predecessors, including the Greek, Roman, and Byzantine cultures.

The globe has not survived the centuries in perfect shape. The 1.5-millimeter wall of the sphere is not complete; there are faint cracks, and a small hole obscures part of the constellation known as Serpens. A few of the constellations are missing a star, and a few have too many. The globe is unsigned and careful scholarship suggests that this is probably because it was likely abandoned by its maker when a major and uncorrectable projection error was noticed.

Globes of this type do not distort the visible characteristics of the sky, or the systems of coordinates astronomers have created to map the heavens. The 12 signs of the

so-called zodiac are clearly visible on this globe, engraved with meticulous care to represent the mythological figures that divided the sky roughly into months, as measured by the sun's apparent motion among the stars during the course of the year. The word *zodiac* itself is derived from the Greek for "carved animals." The 12 "signs," or constellations, date from antiquity, but their spe-

cific outlines and extent vary from era to era. Astrological and calendrical formalisms divided the zodiac into 30-degree wedges to represent what were the standard signs, whereas the orbit of the moon was divided into some 28 stations called "lunar mansions," each representing the position of the moon on a particular day.

—*David DeVorkin*

The heavens—as seen from outside the starry sphere—adorn this celestial globe, which dates from about 1630. Celestial coordinates and some 48 constellations are engraved on its surface. Bits of silver mark the positions of about 1,000 stars.

STAR MAPS: A CONFLUENCE OF ART AND ASTRONOMY

the same ones Ptolemy described, and the stars in each constellation are numbered consecutively, the Arabic numerals corresponding with positions in Ptolemy's catalog. Also following Ptolemy, the planispheres show the celestial sphere from the outside, with the constellation figures facing in toward Earth, and seen from the rear. Because we know how the maps were drawn—from the projection of the sky onto a flat surface—we can reconstruct the positions of the stars they show. Having done this, we know that, for some 90 percent of the stars, the latitudes agree to within 1 degree of those given by Ptolemy, as do the longitudes. Some 125 stars, however, are badly askew. While these erroneous positions might indicate carelessness on Heinfogel's part, it is more likely that he copied these positions correctly from an incorrect catalog. The catalog he used has not yet been identified, but we have found two manuscript planispheres with similar errors, so there is some basis for this theory.

A 13th-century monk measures the height of a celestial object with an astrolabe, while two assistants record his observations.

The date of the Nuremberg planispheres presents a similar problem. When analyzed from a modern point of view, with a modern understanding of the periodic variations that occur in stellar positions in the sky as the result of precession and other forces, the stellar longitudes appear correct for around A.D. 1440. But when analyzed from the scientific perspective prevalent in Nuremberg at the time these maps were made, these longitudes are correct for around 1500. This suggests that Heinfogel was probably strongly influenced by the Alfonsine Astronomical Tables—the best and most widely used star catalog in Europe in the early 15th century. In 1251, when the Tables were compiled, the Alfonsine star positions were relatively accurate, but because of the Alfonsine's quirky theory of precession, the errors in stellar longitudes increased fairly quickly. By 1500 these positions were about 1 degree too small, thus accounting for the error found on the Nuremberg planispheres.

There is at least one other unsolved problem of the Nuremberg planispheres, which is that these are the first star maps on which the 12 zodiacal signs are identified by their symbols. The origin of these symbols is unknown, and their early history is vague. The symbols would, however, show up on many subsequent maps.

78

In Belgium in 1551, Flemish cartographer and geographer Gerard Mercator produced the first celestial globe covered with tapered printed paper panels. This "published" celestial globe measured about 16 inches in diameter, and was the mate to a globe of the Earth that Mercator had published ten years earlier. Mercator's celestial globe is basically Ptolemaic, but with star positions correct for 1550. The stellar longitudes used on the globe were probably derived from the observations that Bernard Walther had made in the late 15th century, after German mathematician and astronomer Regiomontanus had recognized some of the accumulated errors in the Alfonsine Tables. Because Walther worked in Nuremberg, it is surprising that the Nuremberg planispheres were not informed by his work. These new longitudes were also published in Johannes Schöner's *Globi Stelliferi, sive Sphaerae Stellatum* in 1551.

The Mercator globe depicts the 48 Ptolemaic constellations and two smaller star patterns, or asterisms, that Ptolemy had mentioned—Antinous and Cincinnus. Antinous surrounds the six "unformed" stars of Aquila the Eagle—that is, the six stars that Ptolemy included in the *constellation* of Aquila, but not in the *figure* of Aquila. Cincinnus, a beautiful semicircle of stars (later known as Coma Berenices) surrounds three "unformed" stars of Leo the Lion. Antinous remained an independent figure until the early 19th century, when its stars were again included in Aquila. Coma Berenices is still a constellation.

In the final quarter of the 16th century, Danish astronomer Tycho Brahe redetermined the positions of many Ptolemaic stars and charted some stars that Ptolemy had overlooked. Dutch cartographer Willem Janszoon Blaeu visited Tycho at his observatory on the island of Hven in the winter of 1595–96 and obtained a copy of his new but as yet unpublished star catalog. Returning to Amsterdam, Blaeu published a celestial globe based on Tycho's work. On it, Blaeu also depicted the nova in Cassiopeia, a bright star that Tycho and others had observed in 1572. The discovery of the nova had led to a heated discussion of whether celestial objects were as immutable as previously believed. Blaeu's globe was probably issued in 1598; the star positions are said to be correct for 1600.

Although easily recognizable, Blaeu's constellation figures are stylistically quite different from those shown by Dürer or Mercator. Some reflect contemporary culture. For instance, Argo is represented as a Dutch ship of the period, and Boötes is dressed for a northern European winter. Others seem to have been copied from the Farnese globe. Blaeu probably never saw the globe itself, but only the drawings of it that circulated in artistic circles.

In the 16th century, as northern sailors began exploring the southern seas, they began identifying stars in the southern skies that Ptolemy had not been able to see and therefore did not chart. Blaeu knew about these southern stars, but decided not to include them on his globe, as the accuracy of their positions was substantially inferior to Tycho's northern star positions. But he soon changed his mind. In the 1603 revised edition of his globe, Blaeu included the two new southern constellations formed of Ptolemaic stars, Columba and the Southern Cross (which he called El Cruzero Hispanis), as well as the 196 southern stars and 12 southern constellations that Dutch navigators had recently charted. These Dutch constellations included Apus, Chamaeleon, Dorado, Grus, Hydrus, Indus, Musca, Pavo, Phoenix, Piscis Volucris, Triangulum Australe, and Toucan. Around that same time, Johann Bayer, a lawyer in Augsburg, Germany, produced the first important celestial atlas, the *Uranometria*. Published in 1603 and based on Tycho Brahe's catalog, the *Uranometria* contains a copper engraved chart of each Ptolemaic constellation, as well as a chart of the new southern stars and two planispheres. On the back of each chart Bayer printed a discussion of the various names for the pictured constellation and a catalog of its stars. This work was immensely influential. The text was reissued five times, and the charts—without the accompanying text—were reissued eight times.

Bayer's most enduring innovation is his method of identifying the stars by letters, Greek for the brighter and Roman for the fainter, with the alphabetical order corresponding, for the most part, with decreasing brightness. Perhaps because of his new method of star identification, Bayer felt free to disregard the Ptolemaic convention of how constellations were to be shown. While Bayer's charts are like Ptolemy's in that they are geocentric, showing the skies as they would be seen from the Earth, they differ in that some figures face the Earth and some face away. As a result, astronomers could no longer easily understand what Ptolemy meant when he described stars as being in the right arm of one constellation, or the left knee of another.

This same problem arose in the *Firmamentum Sobiescianum sive Uranographia*, the beautiful star atlas produced by Johannes Hevelius, of Poland, and published in 1687. Unlike Bayer, Hevelius was a real astronomer. Working in his private observatory in Danzig (now Gdansk), he charted the features of the moon, the paths of comets, and the positions of the stars. His *Prodromus Astronomiae*, the star catalog on which his atlas was based, appeared in 1690. In addition to the traditional Ptolemaic constellations, Hevelius included several new constellations surrounding stars discovered during the course of the 17th century. Hevelius himself introduced nine star groups, most of which are still in use.

One of these star groups has a most interesting origin. Although Hevelius is famous for producing some of the most ingenious "aerial telescopes" of the late 17th century—huge, tubeless telescopes sometimes hundreds of feet long and operated by complex rigging like a sail on a sailing ship—he was not a great fan of telescopic sights, or optical telescopes attached to traditional measuring instruments to increase their precision. The images that the simple optical systems of his day produced were fraught with error and could dance around in the telescopic field depending upon where the observer's eye rested near the small eye lens. His aerial telescopes required teams of assistants to manage and elaborate hoisting mechanisms, and yet they worked hardly any better. Hevelius trusted his eyes over telescopic sights and let the world know it. One of his more memorable reminders was a new constellation he called Lynx. To see the stars in this constellation, Hevelius wrote, you must be as sharp-sighted as a lynx.

Among the other constellations that Hevelius introduced was Scutum Sobiescianum, which he named in honor of the king of Poland, John Sobiesci III. Such pandering to political authorities was not unusual at that time. For instance, Edmund Halley devised the constellation of Robur Carolinum shortly after the Restoration of the monarchy in England, its name and form representing the royal oak at Boscobel that hid King Charles II from Republican soldiers. Hevelius included this constellation in his star catalog as well.

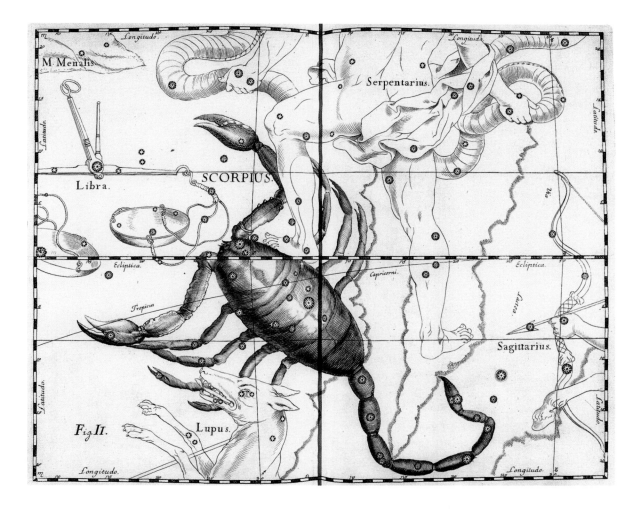

John Flamsteed was another prominent astronomer who contributed to the evolution of celestial maps. He was appointed England's first Astronomer Royal in 1675, and spent much of the rest of his life producing a star catalog and set of charts based on telescopic observations. As an observer, Flamsteed clearly understood that an astronomer working in a modern and well-funded observatory—such as those at Paris and Greenwich—could probably locate a celestial object from knowing its coordinates, but that most rank-and-file observers at smaller installations and outposts could not. Most observers, he knew, described the positions of planets, comets, or new stars by reference to other well-known and easily located objects. Accordingly, he described his charts as "the glory of the work and, next [to] the catalogue, the usefullest part of it." For this same reason, Flamsteed criticized Bayer and other cartographers who neglected the Ptolemaic conventions concerning star positions within the

Johannes Hevelius published an exquisitely rendered star atlas in 1687 based on a star catalog he compiled himself. In addition to the traditional constellations passed down from the ancient Greeks (including Scorpius, shown here), he introduced a few of his own, most of which are still in use.

constellations, and he made sure that his own charts were correct by Ptolemaic standards. Therefore, while Flamsteed's charts are geocentric, his constellation figures all face in toward Earth.

Flamsteed's reluctance to publish his star catalog without the accompanying atlas led to a well-publicized and acrimonious dispute with Isaac Newton, a mathematician who needed the star positions, but who had little appreciation for the pictorial charts. Flamsteed won in the end. His catalog was published posthumously in 1725 as the *Historia Coelestis Britannicae*, while the *Atlas Coelestis* appeared in 1729.

The *Historia Coelestis Britannicae* was the last great celestial atlas. In time, as telescopes proliferated and became more powerful, the number of known stars increased dramatically, and the number of constellations became unmanageable. So, at the same time that technology freed astronomers from their dependence on the visual spatial relationships of traditional star charts, the information to be charted overwhelmed the traditional chart format. By the mid-19th century, sensible astronomers called for reform. They reduced the number of constellations in use, rationalized their boundaries, and eventually omitted the constellation figures altogether. The result was maps devoid of charm or beauty, serving simply as functional tools of modern science.

Today, although professional astronomers have moved away from the colorful and fanciful star charts of the past, these magnificent documents of art and science are still preserved and venerated as relics of the visual era in astronomy. They adorn walls and entryways in museums, and have been reproduced in jewelry and other decorative arts. Some of the most colorful and complex of these, especially those by a late 17th-century cosmographer and rector of the Latin School at Hoorn in the Hague, remain familiar images today. Andreas Cellarius's *Harmonia Macrocosmica*, first published in 1660 in Amsterdam as a supplement to his *Atlas Novus*, had taken him more than ten years to compile. Its 30 magnificent double-folio color plates of the Ptolemaic, Copernican, Aratian, and Tychonic world systems have become icons of astronomical history. The plates survive to this day in many reprinted forms as illustrations in textbooks and magazines, posters, calendars, and even as jigsaw puzzles. They have been reprinted equally by museums of art and science worldwide and have become indelibly linked with the traditional finite universe of the Renaissance.

First published in 1660, the lavishly detailed and colorful images produced by celestial cartographer Andreas Cellarius remain among the most familiar and best loved depictions of the heavens. This star chart, superimposed on a terrestrial map, provides a view of the celestial sphere from the outside looking in.

STAR MAPS: A CONFLUENCE OF ART AND ASTRONOMY

THE GREAT
COPERNICUS CHASE

Owen Gingerich

"Who was Copernicus, and why is his book so valuable?"

It was a warm June afternoon in Federal District Court in Washington, D.C., when the U.S. prosecutor posed his question. I was in an unaccustomed spot, on the witness stand before a jury.

Normally when a thief is caught with a valuable book, he plea bargains, and the book is silently returned to the library from which it was taken. This case was different. The defendant held a security clearance and a plea bargain would cost him his job. But he had been very clever—he had kept the book for seven years before he attempted to sell it, thus allowing the statute of limitations for a civil offense to run out. One small oversight tripped him up. By taking the stolen copy of Copernicus's

As a student in Cracow, Poland, at the end of the 1400s, Nicolaus Copernicus may have used astronomical instruments like these. As Columbus set sail on a voyage that would begin to change people's view of the world, Copernicus was embarking on an intellectual journey that would change people's view of the universe.

De revolutionibus from his home in Maryland to a book dealer in the nation's capital, where he had consigned it with a price of $7,500, he had "knowingly carried stolen property worth more than $5,000 across a state line," thereby setting the clock running again.

As the government's lead witness, I was eager to educate the jury about the book and its author. "Copernicus was a student in Poland when Columbus discovered America," I said. "He studied church law and medicine in Italy, but he was especially interested in astronomy. At that time everyone believed that the Earth was the center of the universe and that every day the sun spun around the fixed Earth, but Copernicus began to think otherwise. Eventually he wrote a book expounding his new blueprint for the cosmos—a sun-centered, or heliocentric, system. The sun, 'as if seated on a royal throne,' governed the planets, circling around it, including the Earth. This was the birth of modern science, a magnificent intellectual leap forward."

In the 1500s Copernicus described a universe in which the sun, not the Earth, lay at the center. Contrary to common sense, this idea represented a daring leap of imagination.

I was bursting with information about Copernicus when the federal prosecutor interrupted to present Exhibit A, a copy of the second edition of Copernicus's book, published in Basel, Switzerland, in 1566. Had I seen this book before?

I studied it carefully, not letting on that the FBI had renewed my acquaintance with the book a few hours earlier. On the stand I consulted my file of Copernican book descriptions.

"Yes, I saw this book about ten years ago in Philadelphia," I said, "but when I later went back to examine it more closely, it was missing." Then I explained how I recognized it. "In the 16th century, books were sold unbound just as a stack of paper, and each owner chose a binding to fit his own taste and pocketbook. This one, with a bright patterned paper cover, is very distinctive. You can also see inside the front cover the glue marks where two bookplates have been removed, one vertical and one horizontal. That combination is quite uncommon, but I just happen to have with me examples of the two different bookplates from the Franklin Institute in Philadelphia."

With a dramatic flourish I produced the bookplates and showed how they fit on the gummed areas like keys in a lock. The prosecutor carried the book and the bookplates to the jury box. The jurors seemed convinced that the book had been properly identified.

The crime would not be a felony unless the book was valued at $5,000 or more. But when the government's prosecutor asked me how much the book was worth, the lawyer for the defense leaped to his feet to object that I was unqualified to answer that question. The judge, no doubt curious what my response would be, overruled the objection, so I cited several auction records where copies of Copernicus's book had fetched well over $5,000.

In the cross-examination, the defendant's lawyer had a particularly smug expression as he asked, "You didn't give all the auction records, did you? What about the copy sold in London on November 12, 1975, for only £2,400 or the copy sold on October 25, 1976, for only £1,000?" Clearly he had done his homework, but I was well prepared.

"Sometimes at an auction," I replied, "a successful bidder can get a real bargain. I recently saw the first copy you mentioned, its margins dotted with fascinating contemporary annotations, and no dealer in his right mind would sell it for less than $5,000. I've seen the other copy, too, and it was defective. That always makes a disproportionate drop in its value."

By now somewhat desperate, the lawyer asked if I had ever taken a course in book appraising.

"No, but I've never taken a course in history of science, yet I'm a Professor of History of Science at Harvard."

"Just answer my question," he snapped, to which the judge retorted in a stage whisper, "He's trying to!"

How did it happen that as an astrophysicist at the Smithsonian Observatory and professor at Harvard, I was on the witness stand? That strange tale is rooted in my passionate curiosity about the nature of science, how science works, and how we validate cosmological claims. How sure are we that the sun is made of hydrogen? Or that the big bang really happened? Or that the Earth is spinning dizzily on its axis, as Copernicus claimed?

For many years I worked with the largest, fastest computer in New England. Leased for the Smithsonian's satellite tracking program in the early days of the space age, the computer provided an opportunity for some of us lucky astrophysicists to explore the structure and composition of the sun and stars when the IBM 7094 wasn't churning out the positions of American and Soviet spacecraft. Working at the cutting edge of research, I soon ran into the problem of whose data to trust when the observations disagreed. Should I accept the Soviet measurements of the intensity of the sun's violet spectrum, or those from the team at Heidelberg, whose results contradicted the Soviet observations? What to believe? Questions like these strike at the very heart of how science works.

Presently I noticed that the 17th-century astronomer Johannes Kepler had faced similar problems in what he called "the war on Mars." Armed with the best information available at the time concerning the positions of Mars, based on the observations of his mentor, Tycho Brahe, Kepler nevertheless found a frustrating level of ambiguity. Determining the path of Mars was a matter of "votes and ballots" as he put it. "If you find this tedious," wrote Kepler, "take pity on me, for I carried out 70 trials." I used the Smithsonian's computer to recalculate some of Kepler's early work. The computer world loved my results: The IBM 7094 solved his problem in eight seconds.

Gradually I realized that historical examples offered special perspective on the nature of science, and the case of Copernicus began to loom large as astronomers prepared for the Copernican quinquecentennial. Nicolaus Copernicus was born in Torun, Poland, in February of 1473. The world had hardly ever had an opportunity to celebrate a 500th anniversary of a scientist, so the 1973 occasion generated international enthusiasm and interest.

Johannes Kepler sought to discover the geometrical design underlying the Copernican cosmos by modeling the universe as a series of nested solid shapes. His laws of planetary motion proved to be a far greater contribution to science.

But the forthcoming anniversary had an intimidating aspect. Copernicus's work had been studied for centuries, so what could I possibly contribute that was new? The answer came in an unexpected fashion and from an unexpected source, a book by British writer Arthur Koestler.

In 1959 Koestler had published a lively but controversial psychological analysis of scientific creativity, *The Sleepwalkers*. Like many successful novelists, he painted his world in terms of good guys and bad guys. Kepler was his hero, while Galileo and Copernicus suffered from his darts. He branded *De revolutionibus* "the book that nobody read," and for a while his slight seemed reasonable. After all, only the first 5 percent of Copernicus's book dealt with his radical sun-centered cosmology, while the remainder was intensely technical, not suitable for bedtime reading except as a cure for insomnia. But a little reflection should have shown that Koestler's epithet was faulty. After all, an "all-time worst-seller" (another of Koestler's descriptors) would hardly have engendered a second edition before the century was out.

In 1970, with the forthcoming Copernican celebration as well as Koestler's characterizations in mind, I had a Smithsonian sabbatical leave, part of which I spent in England. That November I took my family to Scotland, and en route I visited another science historian. We soon fell into conversation about the reputed low readership of *De revolutionibus,* and I suggested that there are probably more people alive today who have read Copernicus's book than in all of the 1500s. In fact, we believed that we could count on the fingers of two hands the 16th-century readers who got all the way to the end of the book: Johannes Kepler, of course; his teacher, Michael Maestlin; Tycho Brahe, Kepler's mentor; Christopher Clavius, the Rome astronomer who spearheaded the Gregorian calendar reform (the calendar we still use); Andreas Osiander, the scholarly proofreader employed by the printer; Georg Joachim Rheticus, the young Wittenberg astronomer who came to Poland to persuade Copernicus to publish his book; his senior colleague who stayed home, Erasmus Reinhold; and maybe Thomas Digges, the first Copernican in England, who published a short English translation of the soaring cosmological chapter of the book. Not even Galileo made the short list—he was never interested in the picky details that made up the bulk of *De revolutionibus*.

Then, two days later, we arrived in Edinburgh and I went to visit the famous library of rare books at the Royal Observatory. Imagine my surprise when I saw their copy of the first edition of Copernicus's 1543 classic: Its margins were brimming with erudite notes and corrections! If thorough readers were so rare, how come the very next copy I saw was brilliantly annotated from beginning to end? It took a little while to sort this out. Like an Agatha Christie mystery, a cast of suspects had been introduced, but could it really be so simple that the anonymous annotator was already on the short list?

The answer, amazingly, was yes, although it took some detective work to establish his identity. Hardly a household name, Erasmus Reinhold was the Wittenberg professor whose Copernican tables helped make Copernicus famous. The most fascinating aspect of Reinhold's reading was epitomized by the Latin motto he had written on the title page: "Celestial motion is uniform and circular or composed of uniform and circular parts." Missing was any comment such as, "This author stops the sun and throws the Earth into dizzying motion!" What we find so wonderfully memorable about the book, the reason a first edition fetches half a million dollars at auction today, is the revolutionary notion of heliocentric cosmology that Copernicus puts forth in it, but Reinhold managed to ignore that completely. Instead, he concentrated on another aspect of Copernicus's work—his insistence that celestial bodies move in circles, today an outmoded notion that went nowhere, a veritable dead end.

How can we understand this colossal misjudgment on the part of the leading astronomical teacher in the generation following Copernicus? And does his mistake cast a penetrating spotlight on the early reception of Copernicus's ideas?

First, we must understand that Copernicus had no proof that Earth moved. It was merely "a theory pleasing to the mind." It was also completely contrary to common sense. If the Earth was really spinning, wouldn't a stone thrown into the air land in another county? Not only did Copernicus's theory threaten the long-established Aristotelian physics, but it was on dangerous ground with respect to holy scripture. After all, Psalm 104 says, "The Lord God laid the foundation of the earth that it not be moved forever." It was Kepler who later pointed out that the Psalmist was merely praising God for creating a stable environment where life was possible, and that the passage had nothing to do with cosmology, but his ideas were not in tune with the developing Biblical literalism of those times (or even today).

Sixteenth-century astronomers, led by Erasmus Reinhold, found Copernicus's ideas about "uniform and circular motion" technically ingenious, congenial to their tastes, and nonthreatening to their traditional ideas. As for the unproven conjectures about a moving

TYCHO BRAHE: MASTER BUILDER AND OBSERVER

"Amazed, and as if astonished and stupefied, I stood still...with my eyes fixed intently upon it.... When I had satisfied myself that no star of that kind had ever shone forth before, I was led into such perplexity by the unbelievability of the thing that I began to doubt the faith of my own eyes."

Thus did Danish astronomer Tycho Brahe speak of the bright new star that appeared in the sky in 1572 above his home and workplace, Herreved Abbey. He took careful measurements of its position over time, and realized that the distance to the star was greater than that to the moon. In doing so, he demonstrated that the heavens were not changeless. Although Tycho was never a Copernican, it was his observations that brought down the Aristotelian universe.

Observing the heavens was Tycho's passion, and precision was his obsession. He came from aristocracy, so in 1576 King Frederick II of Denmark granted him control of a small island called Hven, located off the Danish coast. With continuing support, Tycho obtained greater levels of patronage and built two major observatories on his island, filling them with the finest instruments, many of which he designed himself. He cataloged the positions of a thousand stars and tracked the motions of the sun, moon, and planets. His accuracy remained unsurpassed until the invention of the telescope.

His observations of a comet in 1577 proved that comets moved about freely through the realm of the planets, a discovery that shattered the centuries-old notion of solid, transparent heavenly spheres. Both of these discoveries convinced Tycho that the Aristotelian universe model had to be modified.

Tycho's first observatory was also his castle, which he named Uraniborg, or "Heavenly Castle." Four great conical roofs marked Uraniborg's observing rooms. The complex also contained a library, an alchemical laboratory, facilities for designing and building instruments, and even a printing press. A few dozen yards away was Tycho's second observatory, called Stjerneborg, where even larger instruments were installed.

Tycho had many assistants, first at Hven and later in Prague, where he moved after his Danish patronage ended. One of his assistants at Prague was the outspoken Johannes Kepler, who had written a powerful defense of the Copernican system, the *Mysterium Cosmographicum*, in 1596. Although Tycho was not a Copernican, he recognized Kepler's talents and hired him as a mathematician to compute planetary orbits. Kepler became the Imperial Mathematician to Emperor Rudolph II after Tycho died in 1601, and used the best of his collected data on the motion of Mars to show that planets did not travel in circular orbits, but in ellipses, banishing the last remaining tenet of Aristotelian cosmology.

—*David DeVorkin*

In the foreground of this image, astronomers are making observations with Tycho Brahe's "Great Mural Quadrant," one of his most accurate instruments. Above the quadrant's arc is a fresco of Tycho directing other assistants in his observatory.

Earth, they preferred to suspend judgment, just as a scientist today might do in the face of beautiful but apparently wild ideas about the cosmos, such as inflation, the proposal that the universe momentarily expanded at a rate vastly greater than that observed today.

The remarkable content of the annotations in Erasmus Reinhold's copy of Copernicus's book gave me an idea for an anniversary research project. If this one copy could offer so much insight into the reception of the Polish astronomer's ideas, what might I find in other copies of the book? Little did I guess that this innocent-sounding quest would lead far beyond 1973, not only to Federal District Court in Washington (where the thief was convicted and lost his job), but to a much better understanding of how scientists communicated in the late 1500s and of the extent of the Roman Inquisition's censorship of science in the days of Galileo.

Finding copies of Copernicus's book was easy at first because there are many search aids such as the National Union Catalog (NUC) at the Library of Congress, but my goal gradually grew more ambitious: to locate and examine every possible surviving example. I made many inquiries, and I depended on a network of librarians, book dealers, and private collectors to alert me to the existence of additional copies in unexpected places. I eventually found out, for example, that the NUC lists fewer than a third of the copies in the United States. Ultimately my worldwide census described the history and state of annotations in 277 copies of the first edition of *De revolutionibus* and 324 of the second.

One of the fascinating discoveries the census revealed is that important annotations are rarely unique. Reinhold's students transcribed his marginal remarks into their own copies of the book, and their students in turn copied those copies. All together, over a dozen copies of the book show parts of Reinhold's extensive marginalia. A far-flung network of 16th-century astronomers pondered Reinhold's clarifications and corrections of the printed text. It took the better part of a decade to prove Koestler wrong in his claim that *De revolutionibus* was the book nobody read, but the bottom line is that nearly every astronomer who took himself seriously in that century owned and annotated Copernicus's book.

The most common of the pen-and-ink manuscript annotations in the book are those set forth by a Papal decree in 1620, following on the heels of Galileo's attempts a few years earlier to sell to the Roman hierarchy the idea of a free and open discussion of the heliocentric system. Fearful of an unproved system that seemed to contradict a literal interpretation of the Bible, the Inquisition decided not to ban Copernicus's book outright, but to brand his ideas as simply hypothetical. So sensitive was the issue that the Dominican censors didn't just leave it up to local clerics to correct the book as they did in other cases. Instead, in this one instance, the Inquisition specified ten places where the text was to be amended to make the heliocentric arrangement appear as a hypothesis and not as a real description of the physical world.

Of course, this makes it easy to look at a copy of *De revolutionibus* to see whether or not it was censored. What I found was that about 60 percent of the copies in Italy were "corrected," and virtually none outside of Italy were touched by a censor's hand even though many of the books were in Catholic libraries. Apparently ecclesiastical authorities elsewhere decided the Galileo affair was a local Italian imbroglio, and they were having no part in it.

In the three-decade-long quest to examine the copies of Copernicus's book, a particularly revealing adventure took place just as the Soviet Union was crumbling. I had acciden-

tally learned—through a mistake on a Russian librarian's part, as it turned out—that the Lenin State Library in Moscow held six copies of Copernicus's book, three of the first edition and three of the second. But try as hard as I could, it seemed I could never examine the sixth copy.

One of the five copies I had been able to see had an interesting source: An inscription in the front stated that it had been presented by the rector of the Halle University in East Germany to the Russian general of the "liberating army." I later discovered that Halle University had another copy, but when I went to see it, behind the iron curtain, a stubborn functionary blocked the way. First, he insisted on giving me a lecture, in German, on communism. Then he added that some of the older professors still believed in God, but the younger ones were more enlightened. "What do you need to know about the book that you can't learn if I look at the book and tell you?" he wanted to know. In the strained conversation, which stretched

Author, historical detective, and one-time witness for the prosecution, Owen Gingerich examines the marginal annotations of the most important first edition of *De revolutionibus* in America at Yale's Beinecke Library. Beyond that is his own copy of the second edition, also notable for its marginal annotations. In *De revolutionibus*, Copernicus introduced his theory of a sun-centered universe.

THE GREAT COPERNICUS CHASE

my spoken German to the limit, I finally caught on that he was extremely resistant to showing me the book because one copy was missing.

"Oh," I said, "I know where your missing book is. It's in the Lenin State Library, the gift of your rector to the liberating army."

"Ausgeschlossen," was his reply. "Completely out of the question. The rector wouldn't have had any right to give the book away without the approval of the senate."

"It's in Moscow," I retorted, "and it's bound together with Stadius's *Tabulae Bergenses*."

He snorted and dashed off to the card catalog to see if the missing copy was bound with a book by Johannes Stadius. A short while later, he returned, completely shaken. It was as if his confidence in the entire system had just collapsed. He led me into the reading room, showed me the book, meanwhile breaking every library rule that he had earlier forced me to read. My wife, Miriam, the silent observer of this episode, could hardly contain her amusement, especially as she noticed the expressions on the faces of the other library workers behind the gatekeeper's back.

Galileo trained his telescope on the sky in the early 1600s and turned the universe upside down. His observations led to the acceptance of the idea of a sun-centered universe.

Meanwhile, my attempts to survey the sixth copy in the Lenin State Library went nowhere. I was told it was "in conservation" and unavailable, and each time I requested a microfilm, they sent me a reel with a copy I had already inspected. Finally, Miriam said, "Look at this article where an American scholar describes all the excuses the Russians give when they don't want you to see something. It's the same thing that is happening to you!"

So I had to face the possibility that my census would have to omit the description of a copy known to be rather heavily annotated. There the matter stood until the winter of 1989–90. Then, some months after the Berlin Wall fell and the Soviet empire was collapsing, I got a telegram from Moscow. "If you want to see the sixth Copernicus, come now. You will be our guest while in Russia."

Extravagant as it seemed to fly to Moscow to inspect one copy of Copernicus's book, it also appeared to be an especially interesting time to see what was going on in the former Soviet Union. Just as I was booking my flights, war broke out in the Persian Gulf, and our Smithsonian travel office begged me not to go, but security on the planes had never been tighter, so it was actually a propitious time to fly.

Within 24 hours of my arrival in Moscow I had seen and photographed the Copernicus volume, a particularly interesting second edition well annotated by Herwart von Hohen-

94

burg, the chancellor of Bavaria and a frequent correspondent with Johannes Kepler. But one thing struck me as odd: Clearly the book had never needed, nor ever had, any conservation. When the young postdoctoral student who had taken me to the library remarked that he had been there the previous day to make sure the book would be available, and that he had been quizzed as to whether I *really* had permission to see the book, I realized that something fishy was going on. I finally got the explanation from my host. After the Second World War the Soviets captured truckloads of books from East German libraries in reparation for their own tremendous losses. However, by the Geneva Convention, these could be considered cultural treasures and hence subject to return. In particular, the Copernicus book had come from the Leopoldiana, a venerable natural history academy in Halle. I duly reported this to the Librarian of Congress, a specialist in Russian studies, who allowed that the experts had always suspected this, but my report was the first actual evidence.

A happier adventure spun itself out over a quarter of a century of investigating copies of *De revolutionibus*. In 1974 I undertook a field expedition to a series of provincial libraries in Italy, including the Biblioteca Palatina in Parma. There I saw a second edition of the book, but the library's first edition was nowhere to be found. The librarian even showed me the gap on the shelf where it was supposed to be. I presumed that the volume was simply in use in some librarian's office and gave it no further thought until I got home and happened to mention it to one of my fellow Copernicanists, who remarked that, curiously enough, the book had been missing when he had tried to see it a year earlier. This information rekindled my interest in the book, so I wrote to the library asking for the description of the volume as it appeared in their catalog. Besides the normal bibliographic information, the citation indicated that the volume was prefaced by a Greek manuscript poem in the hand of the great 16th-century scholar Joachim Camerarius.

I knew of one book, undoubtedly the most fabulous copy in private hands, that matched this description. It had turned up, mysteriously, in the London book market some time after World War II. Could that copy have been "liberated" by an Allied soldier, or by a hungry librarian? Because its current owner lived part of the year in Cambridge, I asked him if he would bring his precious book in to the observatory. When he did, we took it to a dark closet and inspected it with ultraviolet light, under which the rag paper used in 16th-century editions glows except where ink has been applied. Even if the ink has been carefully washed out, wherever the ink once appeared the fluorescence of the paper has been destroyed, and the writing will show clearly in contrast to the glowing paper. Similarly, the glue marks from removed labels reveal themselves in tell-tale traces under ultraviolet light, but there were none! The book with the Camerarius manuscript seemed to have a clean bill of health, yet the description in the Palatina catalog appeared to be such a perfect match that I was prepared to link the missing Italian copy with the one that had been brought to my office.

An unexpected episode a few years later taught me to be more cautious in making such inferences. Pierre Berès, an eminent French book dealer, asked to see me, saying he had something of interest concerning Copernicus. When I visited him, Berès related a fascinating tale: An anonymous caller, possibly Italian, had inquired by phone as to whether he was interested in some old books, including a first edition Copernicus. When the book dealer told the caller he was, indeed, interested, the mystery voice said, "You can't contact us, but we'll get in touch with you." Some days later, Berès said, a packet of photocopies arrived without a return address. The sheets were mostly copies of title pages, but with the Copernicus were two photocopies recording a Greek manuscript poem, and these he produced for me. I saw at once that here was another copy of Camerarius's poem.

"En garde!" I said, and then I explained the situation at the Biblioteca Palatina.

"This is surely the stolen Palatina copy," Berès quickly concluded, and he declared that he would abandon the pursuit of the book. With that, the trail went cold, leaving me with a troubling dilemma: In my census should I publish the photocopies to alert possible buyers to a stolen copy, or should I suppress this information lest the thief simply destroy the identifying sheet?

After a decade, and before I had to make a final decision, the book prefaced with the Camerarius poem surfaced once again. This time a New York dealer had caught wind of its existence. A few weeks later I was telephoned by an expert at Sotheby's in Milan. The Copernicus volume, unbound and in a somewhat disreputable state, had been brought to the auction house by a family in Parma. The book had been acquired by the father, since deceased, who had had a reputation as a local bibliophile, and who apparently did not raise questions about the sources of the books offered to him. What evidence did I have, the voice from Milan inquired, that the book belonged to the Palatina? He explained that he had investigated and discovered that most, if not all, of the books being offered were listed in the Palatina catalog, but in fact were missing from that library, strong circumstantial evidence for their true ownership. However, the absence of specific physical evidence created a sticky problem.

There the matter rested for several months. Then, in January 2000, I received an e-mail stating simply, "Nicolaus is back in the Biblioteca Palatina." I'm not sure precisely how that happened, but I know it would never have taken place without my dogged, adventurous, almost quixotic 30-year pursuit into the way 16th-century astronomers reacted to the most revolutionary scientific advance in more than a millennium.

The Earth, with the moon revolving around it, correctly appears
as the third planet from the sun in this lavish depiction of the
Copernican universe by Andreas Cellarius.

THE GREAT COPERNICUS CHASE

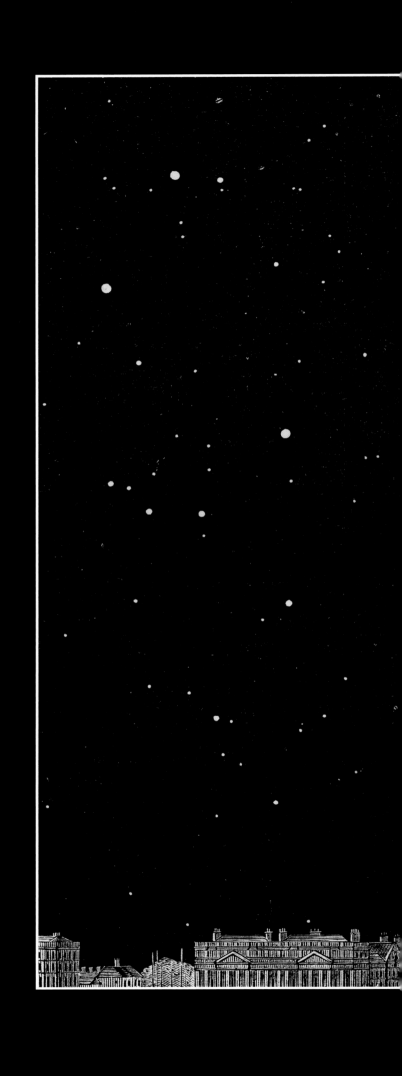

SECTION II

THE
MODERN
UNIVERSE
EMERGES

Telescopic examination of the sky in the 16th–18th
centuries resolved one key mystery of the
universe: The hazy path of the Milky Way clearly
consisted of light from countless stars too dim or
distant to appear as individul points of light.

THE MODERN UNIVERSE EMERGES

David DeVorkin

We launch Section II in the age of Newton and the full expression of the Copernican Revolution. Not only is the Earth displaced from the center of the universe, but the universe itself is now driven by the force of gravity from innumerable centers of mass. We move from a kinematic universe where motions and changes were simply recorded, to a dynamic universe where the motions are understood as the result of forces acting through time. The sun, moon, Earth, and planets all follow the same laws describing these forces at the outset of Michael Hoskin's essay, and by the time he finishes, we find that these forces have been extended to the limits of the visible stellar universe.

The singular device that fostered this extension, and confirmed not only the Copernican model but drove refinements in Newtonian gravitational physics and led to the first evidence that the Earth is in physical motion about the sun, was the telescope. Telescopic observations of the heavens at the beginning of the 17th century, initiated by Galileo, confirmed the idea of a sun-centered universe by showing that many of the beliefs incorporated into the classical Earth-based model were not correct. Telescopes offered a new and powerful way to gauge the universe, increasing our ability to pinpoint the positions and motions of stars and to detect minute variations not sensible to the eye alone. In Galileo's hands, Jupiter became a center of motion for its four starry moons, Venus exhibited phases as it moved around the sun, and the sun itself had imperfections on its surface, which indicated that it rotated on a physical axis. Galileo also found that the starry realm had depth: His telescopes revealed more stars in the Milky Way and in various spaces that heretofore had been empty.

This new telescopic capability greatly stimulated the search for the size of the universe: the distances to the sun and stars. Accompanying this new search was the recognition that our universe is a three-dimensional realm of stars. As Michael Hoskin recounts, based upon the incomparable resources at his disposal in the Newtoniana of the Cambridge University Library, new questions arose about the structure and form of the starry (or sidereal) universe. The stars were no longer fixed on a celestial sphere, the speed of light was no longer infinite but actually measurable, and it was a universe dominated by gravity that could be explored using the "celestial mechanics" that arose from Newton's laws of motion.

Although most astronomers worried about positions and motions of the sun, moon, planets, and stars, a few looked deeper using larger telescopes of a new breed. William Herschel founded modern observational cosmology in the late 18th century when he set about to actually *measure* the scale and structure of the sidereal universe. Herschel was the first observational astronomer to pay more attention to the organization of the sidereal system and the classification of the objects within it than to the motions and appearances of the planets. His study of the enigmatic nebulae led him to speculate on their status as island universes beyond the Milky Way. In Herschel's wake, close examination of the Great Nebula in Orion by an even larger telescope in the mid-19th century appeared to show that it was in fact, a cluster of stars. Hence astronomers remained skeptical of the existence of true nebulae.

Hoskin's chapter ends with Herschel's son, John, still pondering the nature of the enigmatic diffuse nebulae. Robert W. Smith then begins his reconnaissance by exploring what was needed to determine once and for all that the majority of these nebulae were indeed vast Milky Way–size stellar systems. In the late 19th century, spectroscopy demonstrated that true gaseous nebulae existed as vast clouds in the spaces between the stars, and an efficient spectrograph in the hands of Lowell astronomer Vesto Melvin Slipher revealed that the spiral nebulae were traveling at speeds far in excess of any star. But it was not until Edwin Hubble devised a means

By the mid-1600s, Kepler, Galileo, and others had firmly established that the Earth and planets circled the sun. However, the nature and extent of the universe beyond the solar system remained a mystery.

of reliably determining the distances to galaxies that he found a clear relationship between distance and speed: The more distant galaxies were receding faster than the closer galaxies.

One of the central themes of the exhibition is how we removed ourselves out of a static, finite universe of stars and into a dynamic, expanding universe of galaxies. Smith's survey thus begins and ends with a single name, because it was Edwin Hubble's pathbreaking discoveries in the 1920s that stimulated NASA to name its Space Telescope in his memory. In effect Hubble's work has guided cosmology in the 20th century and is likely to continue well into the 21st century. He set much of the intellectual agenda for the Hubble Space Telescope, as it continues to refine his own detection of the rate of expansion of the known universe.

Every bit as profound as the new universe revealed by Hubble's observations and conclusions is the realization, gained in much the same time frame, that the chemistry of the visible universe is far different than had been supposed. It is a universe dominated by the lightest of elements, hydrogen and helium. And it is far different in chemical makeup than the world we know on the surface of the Earth. This new universe was made possible by the application of the spectroscope to astronomy. Spectroscopy revolutionized observational astronomy in the 20th century just as the telescope did in the 17th through the 19th centuries.

FROM NEWTON'S UNIVERSE OF STARS TO HERSCHEL'S NEBULAE

Michael Hoskin

For the historian of science, an appointment to teach at Cambridge University is akin to winning the lottery. The University Library alone houses the scientific papers of Charles Darwin, James Clerk Maxwell, and Isaac Newton, and the individual college libraries are themselves treasure troves. Trinity, for example, has Newton's personal collection of books, while King's has many of Newton's alchemical and religious papers—documents declined by the University Library in the 19th century as not of interest to historians of science!

Such good fortune befell me in 1959. I had opened a history of science book for the first time only two years previously, and my appointment to teach a subject of which I knew next to nothing owed everything to the support of Herbert Butterfield, the

The Milky Way, the bright band of stars visible through the center of this image, remained only of passing interest to most astronomers until William Herschel set out to discover "the construction of the heavens."

historian whose *The Origins of Modern Science* was proving so influential. Butterfield was very familiar with an age when, if a professor of chemistry was needed, it seemed not unreasonable to take a versatile professor of Greek and give him three months to learn the subject, and so he took a chance on me. As a pure mathematician, I knew next to no science, next to no history, and no history of science. My early months in the subject were fraught with difficulty. On occasion I would slow down my delivery in the lecture room not simply because I had not prepared anything more, but because in a few minutes I would have taught them every last thing I knew about the subject.

Because, prior to the development of astrophysics, astronomy was essentially applied geometry, it was on this area that I concentrated my research. A surprising number of my colleagues, presumably of masochistic tendencies, had chosen to bury themselves in the minutiae of planetary theory (in the works of Hipparchus, Ptolemy, Copernicus, Kepler, Newton, and so forth). It was true that complex geometrical models were of primary interest to the astronomers of the distant and not-so-distant past, for whom the stars were little more than an unchanging backdrop to the planetary motions, but planetary theory is a tiny part of modern astronomy, and the emergence of stellar astronomy and of a scientific cosmology was therefore a historical episode of immense importance. By a stroke of good fortune, in the late 1960s Churchill College invited me to take responsibility for the construction and development of an Archives Centre, and this eventually allowed me to offer the Royal Astronomical Society a temporary safe haven for their manuscript collections, which included a wooden chest of papers of William, Caroline, and John Herschel.

Isaac Newton believed that the universe of stars was infinite and symmetrical and that each star was held in place by the gravitational pull of the stars around it.

And so it was that these papers, central to the establishment of stellar astronomy, came into my care, and to my joy they were to remain in the Churchill Archives Centre for decades. Other collections also came into my hands, but the real gold mine was the collection of Newton papers in the University Library, and this is where my story begins.

Isaac Newton was to be revered as the man who had unlocked the primary secret of the physical universe—the inverse-square law of gravitational attraction—but his interest in the universe as a whole was in fact minimal. His masterpiece, *Principia mathematica philosophiae naturalis* (1687), has a great deal to say about the solar system, but it barely mentions the stars.

This is very strange, for ever since antiquity the stars had been known as "fixed" because they maintained their positions relative to each other century after century, in contrast to the handful of "wanderers" or "planets" that moved among the fixed stars. The fixed stars (Newton consistently uses the Latin adjective *fixae* when speaking of them) had often been thought of as visible features on a heavenly sphere. Since the middle of the 17th century it had been recognized that, in fact, the stars were isolated bodies scattered throughout infinite space, yet they were as motionless as ever—or so the evidence suggested. But how could isolated bodies, free to move in response to forces acting upon them, fail to respond to the gravitational pulls of the other bodies in the universe? Could it be that Newton was simply mistaken in claiming that gravity was a universal force?

At some stage of his career, I thought, Newton must surely have faced up to this paradox of the fixity of every star in a universe dominated by gravity; and the most likely time would have been in the aftermath of his famous correspondence with the young theologian Richard Bentley. Some five years after the publication of the *Principia*, Newton received the first of several letters from Bentley, who had preached a series of sermons (in effect, lectures) on the compatibility of science and religion, and who wished to question the renowned author of the *Principia* before finalizing his text for publication. In his replies, Newton is as guarded as ever, but when the exchange ended, the paradox had been put squarely before him. Only the barest hints of his eventual solution got into print and so became familiar to historians, but it occurred to me that at the time of the exchanges with Bentley, Newton had been preparing drafts for a second edition of his book. As it turned out, he was to leave Cambridge in 1696 for a non-academic post in London, and the second edition that eventually appeared was mainly the work of others. But in the University Library in Cambridge were Newton's own manuscript drafts, and perhaps they would contain his solution to the paradox.

This was the good news. The bad news was that the manuscripts were vast in quantity—for Newton was forever drafting and redrafting and he preserved every scrap of paper—and they were in chaos. The only way forward was to go through all the relevant files, examine each page for any reference to the stars, and ask for photocopies. I then spread the sheets out on the floor of a very large room and looked for related pages. If on one page some words had been altered, and on another page the same words appeared in a clean copy, then it would be clear that the former page was an ancestor of the latter. In this way there emerged a succession of drafts of an intended theorem for the second edition, in which Newton

The crude lenses in early telescopes produced blurry images and distorted colors. Newton designed a telescope that used mirrors in place of lenses, an innovation that solved the color distortion problem.

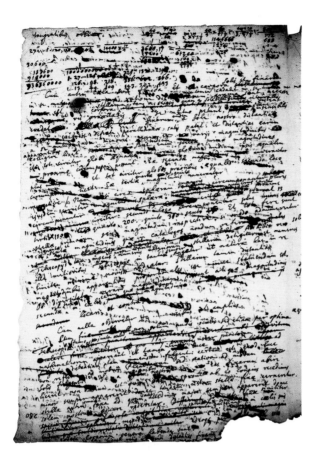
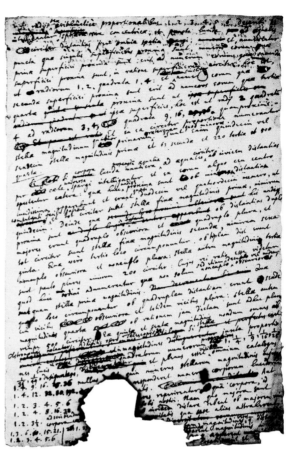

planned to present his solution to the paradox, and indeed his world picture.

Historians were well aware that Newton—a devout if highly unorthodox Christian—believed in the tradition of the two forms of divine revelation, one in the Scriptures and the other in the created universe, and that he saw in the solar system a demonstration of Providence in operation. Newton rejected the conception of Descartes and his followers, whereby God created the universe with its laws of motions, and then left it to run itself without further interference. He believed to the contrary that God not only created the universe as a sublime context for human life, but that he intervened at regular intervals to maintain it as such. He was to be derided by Leibniz for believing in so incompetent a creator as one who had to salvage his creation by the desperate step of miraculous intervention. But for Newton these interventions were not ad hoc miracles, but part of a servicing contract between God and the universe that demonstrated the ongoing concern of the creator for his creation. As for the solar system, Newton believed Providence had created it in a highly stable configuration of near-circular orbits, with the planets Jupiter and Saturn—whose large masses were poten-

The carefully preserved drafts Newton kept of his work provide insight into the evolution of his ideas. On these pages he grappled with the problem of showing how the evidence from star catalogs fit his idea of a symmetrical universe.

tially disruptive—banished to the outer reaches where they could do the least harm. Nevertheless, the time would come when the system would be threatened with instability, and Providence would then intervene and restore the original order.

The draft theorem showed that Newton had a comparable picture of the universe of stars. The stars were clearly not in orbit like the planets, for they were "fixed," and therefore the stellar system could not be finite, because a finite system of stars initially at rest would soon suffer gravitational collapse as all the outer stars were pulled in toward the center. The stellar system was therefore infinite, and, at the beginning, Providence had imposed a high degree of symmetry so that each star was pulled by the other stars equally in opposite directions and therefore continued to be at rest.

But a glance at the night sky shows that the symmetry is imperfect. This is most evident in the Milky Way, but, curiously, the 17th century had little interest in the Milky Way or the shape of the system of stars that formed it. The lack of symmetry is clear, and to Newton this would not have been wholly unwelcome, for it teaches us that we must rely on Providence to intervene at intervals, restoring the stars to their original order to prevent their movements from getting out of hand and leading to chaos.

It was one thing for Newton to conjure up this model of the universe of stars, but quite another thing for him to justify it by testing his model against the evidence. The evidence was contained in star catalogs, and these listed the positions of the stars—that is, their directions from Earth—and their apparent brightnesses. His model was three-dimensional and gave no privileged position to Earth, but the catalogs were two-dimensional, with only brightness as a clue to the third dimension of distance, as seen by Earth-based observers.

To make a test, however questionable, Newton had to derive a third dimension from the brightness, and he therefore assumed that the stars were physically similar and that the brighter of two stars would always be the one nearer the Earth. He had no photometer with which to measure and compare the brightnesses of stars, so he assumed that a star deemed to be of magnitude "n" in the traditional classification was at a distance of "n" units. But this did not remedy the problem that the evidence was Earth-centered. To get around this, Newton replaced his symmetric model with one centered on the solar system, in which the stars were located on imagined spheres concentric with our solar system and of radii 1, 2, 3, (and so on) units. On the surface of the sphere of radius 1 would be the stars of magnitude 1, the brightest, because they were by hypothesis the nearest. How many such stars could be fitted onto the surface of the sphere so that they were not unduly close—that is, no nearer to their neighbors than they were to the solar system? Newton knew the answer to be 12 or 13 (he was not sure which), and, sure enough, this was roughly the number of stars of the first magnitude.

At the next stage were the stars on the surface of the sphere of radius 2. Because this sphere had four times the surface area of the previous one, it had room for four times as many stars, about 50. The next sphere, of radius 3 units, had room for nine times as many stars, and so on. In his first drafts Newton was so confident of a match between his predictions from this model of spheres and the data in the star catalogs that he went ahead on the assumption that all was well, leaving blanks where he would later fill in the actual numbers. But when he eventually did so he found that the numbers of stars increased altogether too rapidly and that something was seriously wrong. What to do? It did not take Newton long to realize that one of his assumptions was expendable. Evidently, the relationship between magnitude n and distance n was not the simple equality that he had assumed, and he was right: Modern astronomy has found it convenient to define a star of traditional magnitude 6 as being at 10 times (rather than 6 times) the distance of a star of magnitude 1. By substituting a more complex relationship, Newton persuaded himself that he had justified his claim that the universe of stars is nearly symmetric.

The drafts of the theorem were to gather dust for three centuries, but we can see from occasional remarks by Newton's intimates that some of them were admitted to his secret cosmology. Not so admitted was William Stukeley (1687–1765), a young physician who became a Fellow of the Royal Society during Newton's long presidency. Newton had focused on the gravitational effect of the stars on each other; Stukeley was interested instead in the appearance of the night sky—that is, on the light sent to us by the stars. In a conversation with the great man he argued that if the system of the stars were infinite, as Newton had discreetly proposed to him as a possibility, then "The whole hemisphere [of the sky] would have had the appearance of that luminous gloom of the milky way."

Edmond Halley (ca 1656–1742) breakfasted with Stukeley and Newton early in 1721, and they discussed astronomical matters. Stukeley must have repeated to Halley what he had earlier said to Newton, because a few days later Halley read to the Royal Society the first of a pair of papers on cosmology, remarking "Another Argument I have heard urged, that if the number of Fixt Stars were more than finite, the whole superficies of their apparent Sphere would be luminous." Halley proposed a solution to the problem; it was fallacious, but his papers were published in *Philosophical Transactions* and for the first time the Newtonian model of an infinite and near-symmetric universe of stars came (anonymously) into the public domain.

A more competent treatment of the question was published in 1744 by Swiss astronomer J.-P. L. de Chéseaux (1718–1751). He pointed out that, in an infinite and regular system of stars, at twice the distance of the nearest stars there was room for four times as many stars but each would appear one-quarter as bright because light diminishes with the square of the distance. Collectively, therefore, the stars at twice the distance would contribute as much to the brightness of the night sky as did the nearest stars, and similarly for those at three times the distance, four times, and so on indefinitely. But did not this imply that the entire sky would be ablaze with starlight? No, because the analysis so far had assumed that the transparency of space was perfect, so that the light of a distant star would reach the Earth-based observer without the slightest loss en route. In the real world this was scarcely possible, and

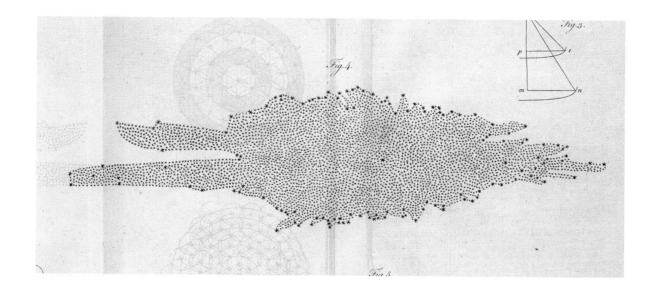

de Chéseaux pointed out that even a tiny loss, repeated over and over again in the successive stages of the journey of distant starlight to Earth, would be sufficient to explain the darkness of the night sky.

A similar analysis was published in 1823 by the German amateur astronomer H. M. W. Olbers (1758–1840), and this time in the widely read *Berliner astronomisches Jahrbuch*. This brought the issue to the attention of the astronomical community, but, before long, developments in physics showed that the absorption of light by an interstellar medium would not resolve the issue, for in time the medium itself would heat up and begin to radiate light. But 19th-century astronomers had no difficulty in finding alternative scenarios. Some suggested that the infinite system of stars might be hierarchically ordered, with disproportionately large increases in the spacings in the upper echelons of the hierarchy; or the stars might be grouped in systems between which there was an etherless vacuum that light was simply unable to pass across. It was only in the later 20th century that cosmologists seized on the darkness of the night sky and made of it "Olbers's Paradox," unaware that in fact it originated with Newton and the musings of his young friend Stukeley.

Stukeley had been only one of a number of 18th-century philosophers and theologians who speculated about the structure of the universe and, in particular, on an explanation of the Milky Way. Another was Thomas Wright of Durham (1711–1786), an itinerant lecturer on a variety of subjects, who achieved some fame as a result of the credit paid him by no less a figure than Immanuel Kant (1724–1804). Wright, it seemed, had been the first person to recognize that the Milky Way is the optical effect of our immersion in a layer of stars.

William Herschel created this cross section of the Milky Way by counting stars with his telescope. He assumed, erroneously, that the stars were evenly spaced and that he could see to the border of the Milky Way in all directions. Still, his was the first map of the three-dimensional structure of the universe based on observational data.

Early in my career, a publisher invited me to write an introduction to a proposed facsimile edition of Wright's *An Original Theory or New Hypothesis of the Universe* (1750), the handsomely illustrated work in which Wright was said to have proposed this model of the universe. But a close reading of the text showed me that Wright had no such conception. Instead of being the first modern cosmologist, he advocated a bizarre universe in which the stars of our system were clustered around one of innumerable Divine Centres.

By a remarkable chance, at about this time a large and disordered bundle of Wright manuscripts was discovered in a drawer, and offered for auction. Under the guise of friendship, I persuaded the dealers to let me sort the papers prior to auction, and, as I worked, a totally unsuspected draft of a later work, *Second Thoughts,* began to emerge. Here Wright proposed that the sky was solid and set with jewels, the "stars" being glimpses of the fiery exterior of the sky: No modern cosmologist here. This led me to examine earlier manuscripts of Wright, which were known to exist but had never been studied. These included a public lecture on God and the universe, in which Wright set about putting the fear of hell into his audience. At the center of the universe, he told them, was heaven, the abode of God; the stars, including our sun, formed a spherical system clustered about this center, and each star was in orbit because otherwise there would be gravitational collapse (and the star would end up in heaven!). Beyond the system of the stars was the outer darkness, hell.

As an experienced popularizer, Wright illustrated his lecture with a vast visual aid that displayed a cross section of the universe passing through both the Divine Centre and our solar system. To bring matters home to his audience, he employed artistic license and showed the planets, the stars, and the Milky Way as they appear from Earth. The Milky Way, he explained, resulted from the combined light of innumerable stars at the limit of our vision, all these stars being neighbors of the sun in our region of the spherical system.

Evidently he later realized (or did someone in the audience object?) that he had illustrated the Milky Way in just one of an infinite number of possible cross sections that would pass through the Divine Centre and the solar system, whereas the true Milky Way is unique. His ingenious response, presented in *An Original Theory*, was to make the spherical shell of stars that included the sun vast in radius and narrow in depth. As a result, the *visible* stars occupied a small segment of the sphere that was not very different from a disk in shape. When we look sideways, so to speak, in the plane of the disk, we see great numbers of stars whose light merges to give the appearance of the Milky Way. But when we look inward, toward the interior of the sphere, or outward, we see only a few nearby (and therefore bright) stars before our gaze extends into empty space. The Milky Way, then, defines our local tangent plane to the spherical shell of stars that surround our local Divine Centre (for Wright now allowed many such Centres and surrounding systems of stars). Observational astronomy tells us about the stars in the nearby segment of the shell, while theology tells us about the shell as a whole and about the supernatural space in its midst.

Wright did indeed offer an alternative model, in which all the stars of our system occupy a space in the form of a flattened ring surrounding the Divine Centre. In this model, the sun and the visible stars formed a disk-shaped aggregate to one side of the ring. But Wright could offer no reason for the creator's selection of this particular plane for the ring rather than another, and for him the alternative model is clearly second best.

Kant learned about Wright's ideas from an extended review that appeared in a Hamburg journal a few months after the appearance of *An Original Theory*. Crucially, the review contained no illustrations, and Kant cannot be faulted for his creative misunderstanding of Wright's ideas. Not thinking for a moment that Wright envisaged numerous Divine Centres, one of which was at the middle of our star system, Kant thought Wright offered two alternatives, both in the natural order—a spherical system of stars or a ring system. Kant saw no reason why the ring should not be a disk that extended without a break from one edge to the opposite one. He knew of observations of the milky patches known as nebulae that showed them to be elliptical, and he believed these to be analogous star systems. A spherical system would always appear from a distance to have a circular outline, but a disk seen slantwise on would appear elliptical. Of Wright's alternative models, therefore, Kant thought the disk to be confirmed by observations, and he therefore adopted this view of the Milky Way. Wright thereby was credited with seeing our galaxy as the disk-shaped aggregate of stars of modern astronomy.

Kant developed a hierarchical theory of the universe with the disk as one stage in the hierarchy, as did the Swiss polymath Johann Heinrich Lambert (1728–1777). But all this was no more than the speculations of philosophers and amateur theologians. Meanwhile, astronomy continued to be preoccupied almost entirely with the solar system. A handful of stars had been seen to fluctuate in brightness, and a tiny number had moved position over decades or centuries so that the epithet "fixed" was no longer valid. Some dozens of nebulae were known, but opinion was divided as to whether these were clouds of luminosity or star clusters disguised by distance. What was needed was an observer bold enough to break the mold and to make stellar astronomy a component of the science. The man who did this came to London in 1757 as a teenage war refugee, and began to earn his bread by copying music. His name in its English form was William Herschel.

Herschel (1738–1822) was born in Hanover, the son of a bandsman who struggled to give his sons an education. His younger sister Caroline (1750–1848) was less fortunate: She learned to read and write, but when she joined her brother in England in 1772, her arithmetic was virtually nonexistent. There survive among the Herschel papers scraps of exercises her brother devised to teach her the most elementary arithmetic and geometry. Nevertheless, as her brother's housekeeper and scribe, Caroline was to play a crucial role in one of the great partnerships of the history of science.

By 1772, Herschel was established as an organist in the fashionable resort of Bath (as a composer he merits an entry in *Grove's Dictionary of Music*), and his wide-ranging mind was becoming obsessed with astronomy. Not knowing that astronomers were supposed to devote themselves to the solar system, he set his heart on discovering what he called "the construction of the heavens." To study distant and therefore faint objects he needed telescopes with large

William Herschel's large 20-foot reflector was an extraordinary affair. To observe with it he stood on a movable balcony near the top of a scaffold so he could look down the 20-foot wooden tube of his 18-inch speculum-mirrored telescope. And always, his sister Caroline was there to assist in whatever needed to be done. On the night of December 31, 1783, when the sky had cleared after a heavy snowfall, William made haste to observe with his still-not-complete telescope. As Caroline recorded in her diary:

"My brother, at the front of the telescope, directed me to make some alteration in the lateral motion, which was done by machinery, on which the point of support of the tube and mirror rested. At each end of the machine or trough was an iron hook, such as butchers use for hanging their joints upon, and having to run in the dark on ground covered a foot deep with melting snow, I fell on one of these hooks, which entered my right leg above the knee. My brother's call, 'Make haste!' I could only answer by a pitiful cry, 'I am hooked!' He and the workmen were instantly with me, but they could not lift me without leaving nearly two ounces of my flesh behind…the remainder of the night was cloudy…."

More typically, Caroline acted as William's recorder. His telescope usually faced south, looking at the meridian, and sat still, letting the rotation of the Earth cause stars to drift through the field of view. He hired an assistant to do the cranking from the ground, so Caroline could sit in the window at their home near Datchet and "write down, and at the same time loudly repeat after me, every thing I required to be written down. In this manner all the descriptions of nebulae and other observations were recorded…."

Caroline would repeat back to Herschel the description while Herschel was still looking at the object. After about six sweeps accomplished in this way, Herschel found that he could predict when stars would enter by using Flamsteed's catalogue, which Caroline could read at her desk at candlelight—and starting in late December 1793 he added a sidereal clock. Also, he found he could calibrate the winding mechanism to determine the difference in altitude between two objects. Ultimately he used an index board behind one of the ropes to produce a fine measure of altitude markings.

Caroline helped in all of these modifications, testing them, recording them, and then acting as William's observing assistant. But she also had her own observing to do. William built her a small comet telescope—a "comet sweeper"—and so, when she was not working with her brother, she was scanning the heavens for new comets. She proved to be an effective comet discoverer. By 1797 she had discovered eight of them.

Caroline minded the heavens, adopting her brother's mission as her own. Thus we end this short reconnaissance by citing the words of British astronomer James South in 1828. After reviewing William's lifework on the construction of the heavens, his creation of huge telescopes and his lifelong research agenda, especially his years of sweeping the heavens for nebulae, South then asked:

"Who participated in his toils? Who braved with him the inclemency of the

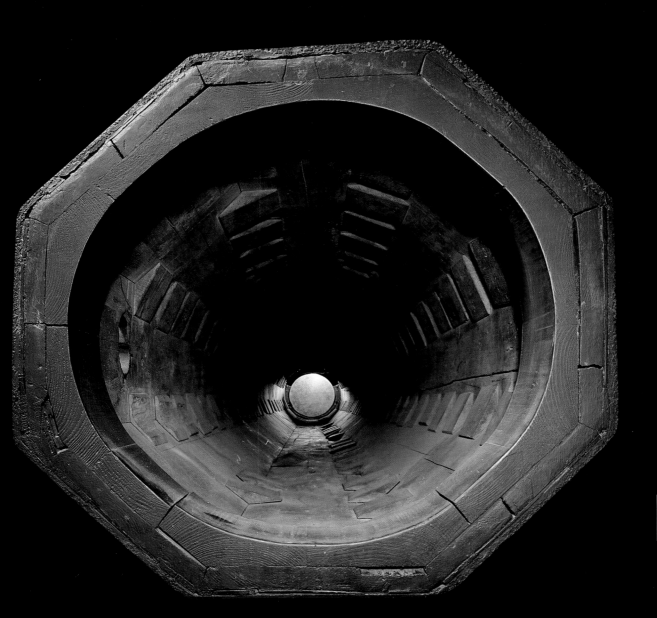

weather? Who shared his privations? A female. Who was she? His sister. Miss Herschel it was who by night acted as his amenuensis: she it was whose pen conveyed to paper his observations as they issued forth from his lips; she it was who noted the right ascensions and polar distances of the objects observed; it was she who, having passed the night near the instrument, took the rough manuscripts to her cottage at the dawn of day and produced a fair copy of the night's work on the following morning; she it was who planned the labour of each succeeding night; she it was who reduced every observation, made every calculation; she it was who arranged everything in systematic order; and she it was who helped him to obtain his imperishable name."

—David DeVorkin

A never before photographed view looking down the tube of William Herschel's 20-foot reflecting telescope toward its 18.5-inch mirror. Herschel began using the telescope in 1783, and although he later built larger ones, this remained his favorite: with it he discovered thousands of nebulae and star clusters.

FROM NEWTON'S UNIVERSE OF STARS TO HERSCHEL'S NEBULAE

mirrors that would collect enough light to make the objects visible to the human eye. These reflectors, he found, he would have to make himself, so he did—casting and polishing the metal mirrors, and devising the mounts.

In the 1770s Herschel was preoccupied with developing his equipment, but he found time to familiarize himself with the brighter stars. It was while so doing, on March 13, 1781, that he chanced upon an object that he—with his homemade telescope and already considerable experience as an observer—instantly recognized as no ordinary star. It turned out to be the planet now known as Uranus, the first planet to be discovered in historic times. Herschel became famous overnight, and his allies at court persuaded King George III to grant him a pension so that he could quit music and devote himself to astronomy.

At that time, Herschel's biggest telescope had mirrors 12 inches in diameter and a 20-foot focal length. What limited its performance was its primitive mounting, for it was slung from a pole and Herschel had to observe from a precarious perch at the top of a ladder. However, his newfound leisure allowed him to devise a stable mounting for a 20-foot reflector with mirrors 18 inches in diameter. It was the best telescope in the world for the study of the construction of the heavens, and, in 1783, Herschel embarked on a task that was to occupy

William Herschel gained worldwide fame in 1781 when he became the first person in recorded history to discover a planet—Uranus. He went on to build the most powerful telescopes in the world and used them to investigate the size and structure of the universe. His sister Caroline assisted him in his observations and telescope building. A skilled astronomer, she discovered eight comets.

his sister and himself for two decades: the systematic examination of the entire accessible sky in the search for nebulae. About 100 were already known, but their nature remained a mystery. Night after night Herschel was at the eyepiece, slowly raising and lowering the great

An engineering marvel, Herschel's 20-foot telescope was the best in the world for studying faint celestial objects. A sturdy framework set on a turntable supported the long telescope tube. To look through the eyepiece at the top of the tube, Herschel devised a platform that could be raised and lowered.

tube while the sky drifted past, with his sister at a nearby desk ready to write down positions and descriptions whenever a nebula passed through the field of view. Together they accumulated data on two and a half thousand nebulae. In the 1830s, after Herschel's death, his son, John (1792–1871), was to take a matching instrument to the Cape of Good Hope and complete the examination of the heavens. John's "Catalogue of Nebulae and Clusters of Stars" is the predecessor of the New General Catalogue that astronomers use to this day.

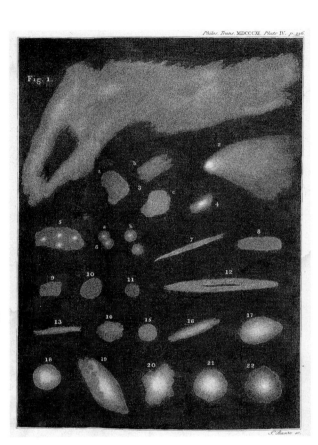

Herschel discovered more than 2,000 nebulae, objects he thought might be clouds of material condensing to form new stars. These reproductions of his drawings show possible stages of nebula evolution.

But what were nebulae? Clearly a star cluster at so great a distance that the individual stars could not be distinguished would appear nebulous, but so would a cloud of luminous matter. At first, Herschel thought that there would be a difference in appearance, that the cluster would display a mottled nebulosity while the cloud would appear milky. But, in 1784, he came across nebulae in which both forms of nebulosity seemed to be present, and he concluded that the nebulae were star clusters in which the milky form was the result of component stars at a great distance, while the stars of the mottled nebulosity were nearer to the observer.

Clusters of stars must be assemblages brought and then held together by their mutual attraction, presumably that of gravity, he thought. And a scattered cluster would, as time passed, become more and more compressed as gravity continued to exercise its grip. It was therefore possible to introduce into astronomy for the first time the concept of objects old and young. The lifespan of the human observer was not long enough for star clusters to form and condense and compress before his own eyes, but Herschel proposed to select from his catalogs examples of clusters at different stages of development, and to lay them out in sequence to represent the life history of a single cluster. It was an astonishing concept.

One night in 1790 he came across a nebula consisting of a single star surrounded by nebulosity, which he instantly interpreted as a star condensing (presumably under gravity) out of a cloud of nebulosity, so he abandoned his identification of nebulae with star clusters: True nebulosity existed after all, and he enlarged his cosmogony to embrace the pre-stellar stage. Physical light assembled under gravity to form clouds of nebulosity, which continued to con-

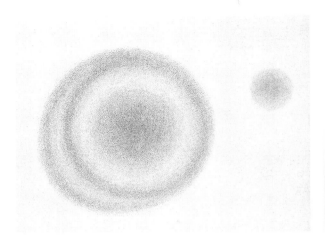

dense until stars began to be formed. As time passed, each of these widespread clusters of stars began to condense until, at length, there was violent gravitational collapse resulting in a dissemination of light, out of which the process would begin again.

I have been reading and writing about William Herschel's life and work for nearly half a century, and I remain as awestruck as ever at his brilliance in introducing into astronomy the methods of natural history, and in replacing the changeless heavens of Aristotle and the clockwork universe of Newton with a cosmos in which heavenly bodies are born, develop, and eventually die. He also fathered another fine scientist, John Herschel, who brought stellar astronomy into the scientific mainstream. Born in 1792, when his father was 53 years of age, John Herschel bore a famous name and enjoyed the best of English educations. Indeed, he was set on a career as a Cambridge don when his father prevailed upon him to become his apprentice in astronomy, charged with the revision and completion of his father's work. William died in 1822, but John had refurbished the decayed 20-foot reflector under his supervision, and Caroline had come out of astronomical retirement to act as amanuensis for John in his first sweeps for nebulae, as she had first done for William nearly four decades before.

In 1825 John embarked on a revision of his father's catalogs of nebulae. As published by William, the catalogs had listed nebulae by type, classified in the manner of the natural historian, but if an astronomer came across a nebula in the sky, it was a great labor to track it down in William's catalogs. John, on the other hand, listed the nebulae by coordinates of right ascension and declination, and made them accessible to the observer. This done—and resolutely declining all offers of government help—he set sail for the Cape of Good Hope where he spent four years extending to the southern skies his father's catalogs of nebulae, star counts, and so forth. John became, and remains, the only observer to have examined (so

By the 1840s others had taken up the quest to build ever larger telescopes. These drawings of nebula M51 testify to the resulting increase in clarity. John Herschel drew the one on the left while using his father's 20-foot telescope. Ireland's Lord Rosse made the other, using a telescope with 16 times more light-gathering power.

to speak) every square inch of sky with a major telescope.

When John set sail for home in March 1838, his career as an observer was ended, and so was the Herschelian monopoly of great telescopes. That year, at Birr Castle in central Ireland, William Parsons (1800–1867), the future Earl of Rosse, assembled a composite mirror three feet in diameter, and the following year he cast a single disk of this size. In 1845 he completed the "Leviathan of Parsonstown" with mirrors no less than six feet in diameter and weighing four tons each. Within weeks the reflector had revealed that some nebulae are spiral in structure.

I was able to examine in person the beautiful sketches made by Rosse (Parsons) and his colleagues when the present Earl invited me to Birr to join in cataloging the scientific papers. His sketch of the Whirlpool nebula, known as Messier 51 and reproduced here (page 117) from a photograph by my colleague Owen Gingerich, has appeared countless times in textbooks and histories over the past 150 years, and gives the impression of a rapidly rotating nebulous mass floating through space.

The Leviathan was designed to settle once and for all the question of whether all nebulae were merely star clusters disguised by distance, and the great nebula in Orion became the decisive test. It happened that the reflector was powerful enough to detect stars that are embedded in this nebula, and Rosse persuaded himself that he had achieved his goal and resolved the nebula into its component stars. Many agreed that this success could be generalized and that belief in "true nebulosity" had been banished from astronomy forever. But in 1864, William Huggins

(1824–1910) turned his telescope onto a nebula in Draco and passed its light through a prism. Instead of seeing the continuous spectrum of starlight, he saw instead the bright spectral lines that showed the light source was gaseous. "True nebulosity" existed after all, and the "new astronomy" of astrophysics had achieved its first breakthrough in the study of the stars.

Nicknamed the "Leviathan," this giant reflector with its six-foot mirror dwarfed the largest of Herschel's telescopes. Built by William Parsons, Ireland's third Earl of Rosse, the Leviathan was cumbersome, but with it Lord Rosse discovered a dramatic spiral structure in M51, known thereafter as the Whirlpool nebula.

FROM NEWTON'S UNIVERSE OF STARS TO HERSCHEL'S NEBULAE

FROM HUBBLE
TO HUBBLE

Robert W. Smith

This is the story of three remarkable telescopes, a faint smudge of light in the constellation of Andromeda, the search for evidence of canals on Mars, and Edwin Hubble, the holder of a bachelor's degree in jurisprudence from the University of Oxford who is now generally considered the leading observational cosmologist of the last century. As a historian of science, I have been fortunate enough to have researched all of these topics, which, in a manner no one could have predicted, became linked in ways that led to profound transformations in the way we study the physical universe.

The Hooker telescope, which sits in a 100-foot dome atop Mount Wilson in California, is one of the reasons for this shift in our understanding of the universe. Many who view it are

Edwin Hubble, seated here at the cameras at Mount Wilson Observatory's Hooker telescope, twice transformed our view of the universe. He proved that most nebulae are actually galaxies lying beyond our Milky Way, and that these galaxies are rushing away from one another. The telescope played a role in both discoveries.

taken aback by its size and bulk, just as I was when I first saw it. The tube of the telescope is mounted in a rectangular steel cradle, and, at its base is a mirror some 100 inches in diameter, its expansive surface designed to collect the light from celestial bodies. It is easy to see from the telescope's appearance that its maker also built ships. Indeed, an uneasy air seems to hang over this leviathan—a giant, yes, but also clumsy, I thought when I first saw it. Yet I knew that such an impression was wildly misleading. When it was completed in 1917 various minor problems still needed to be corrected, but it had already become one of the engineering marvels of the age. It might weigh 100 tons, but it could track with great accuracy stars and galaxies as they wheeled around the sky. The main mirror (or "primary," as astronomers say) was also a triumph for its builders. The painstaking work of grinding and polishing it to an accurate shape had taken many years, unprecedented precautions, and unlimited patience.

I pondered how it might have seemed to a young astronomer named Edwin Hubble when he saw it for the first time in 1919, when it was the world's most powerful telescope. I had spent several years studying Hubble's career, but I had never located an account of his feelings on encountering the telescope, the use of which would do so much to bring him fame, if not fortune. For the discoveries he made, principally with the help of the Hooker telescope, Hubble is now generally regarded as the outstanding observational cosmologist of the 20th century. What, then, had brought Hubble to Mount Wilson in 1919 and what did he do to earn this reputation?

Hubble trained as an undergraduate at the University of Chicago, including a year as a lab assistant to the future Nobel Prize winner Robert Millikan. In 1910 he journeyed to Oxford University as a Rhodes scholar. He emerged with a bachelor's degree in jurisprudence as well as courses in Spanish—not the most likely subjects for a budding astronomer. After Oxford, he returned to the United States and taught high school in Indiana for a year before enrolling as a graduate student at the Yerkes Observatory of the University of Chicago in 1914.

At that time astronomers were unsure if galaxies existed outside of the Milky Way and could be viewed through even the largest and most powerful of telescopes. For more than a hundred years debate had raged over the nature of nebulae, the name then assigned to diffuse milky patches of light seen all over the sky. Two main questions had emerged: Were these dim patches of light made of a

Mount Wilson Observatory's 100-inch Hooker telescope saw first light in 1917 and remained the largest telescope in the world for 30 years. Its main mirror, more than eight feet in diameter, is the largest solid plate-glass mirror ever made.

luminous fluid, or were they remote star systems whose light merged to give a milky effect;? And, if they were groups of stars, were these star systems part of the Milky Way galaxy, or associated with it, or were they themselves vast independent galaxies—"island universes," as some astronomers referred to them? By 1914, the debate was no longer about whether all nebulae were external galaxies. Instead the argument centered on whether one type of nebula—spiral nebulae, (so called because they have a spiral shape like a pinwheel)—were huge star systems far beyond the boundaries of the Milky Way, or whether they were relatively nearby clouds of material within, or bordering, our own galaxy.

Hubble studied nebulae for his Ph.D. dissertation and judged the evidence, on balance, to support the theory that the nebulae were external galaxies, although his writings seem to

A head-on view of the camera used by Edwin Hubble. Eight-by-ten-inch photographic plates, like the one on the facing page, fit into the holder at the center. Hubble peered into the eyepiece just above it and adjusted the wheels around it to keep the plate correctly positioned during exposure.

indicate that he did not regard the evidence as conclusive. Shortly after completing his thesis in 1917, Hubble's studies were suddenly put on hold for two years when he enlisted in the U.S. Army to serve in World War I.

Returning from the service in 1919, Hubble landed a job at the Mount Wilson Observatory, a dream position for an observational astronomer. Here he had the enormous benefits of use of the 100-inch Hooker telescope in his researches and long stretches of time in which to exploit it. Mount Wilson was a private research institution and, for the period, was lavishly funded. George Ellery Hale had founded the observatory in 1905 with monies from one of the great American philanthropies of science, the Carnegie Institution of Washington. At Mount Wilson, Hubble rapidly went beyond the research he had done for his thesis, and was soon tackling the enigmatic spiral nebulae.

The basic question about the spirals that astronomers strove to answer was, "How far away are they?" Various astronomers had calculated the distances using a variety of techniques, but none of these methods were seen as very reliable. One such technique exploited novae—exploding stars that for a brief time rise enormously in brightness but then fade. A few astronomers had detected the flaring of novae in some of the larger, and presumably closer, spirals. If additional novae could be found in spirals, it would be possible, Hubble reckoned, to better compare them with novae that were definitely in our own stellar system. Hubble hoped to secure more accurate distances to the spirals

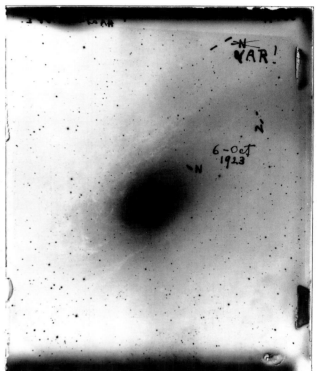

125

Hubble scrawled "VAR!" on this photographic plate to mark the discovery of a Cepheid variable star in the Andromeda nebula. A key to measuring distances, the Cepheid revealed that Andromeda was not a nebula, but a galaxy.

that exhibited novae. In 1923 he turned his attention to the pursuit of novae in a very large spiral in the constellation Andromeda, the Andromeda nebula, which is visible to the naked eye as a tiny milky smear.

Soon he was telling a correspondent, "You will be interested to hear that I have found a Cepheid variable in the Andromeda nebula (M31). I have followed the Nebula this season as closely as the weather permitted and in the last five months have netted nine novae and two variables." The Cepheid variable that Hubble wrote about was a star that exhibited changes in its brightness due to expansions and contractions of its volume. Hubble had found his first Cepheid variable in late 1923, but initially he had mistaken it for a nova. That, after all, was the type of star for which he was searching. But after careful observation he concluded that the star rose rapidly and fell slowly in brightness in a regular manner, consistent with it being a Cepheid rather than an explosive nova. Finding such a star was actually a much bigger breakthrough than the detection of one more nova. Astronomers considered Cepheids excellent indicators of distance, which could be estimated by comparing their luminosity and brightness to locally known Cepheids within the Milky Way galaxy.

In the wake of discovering this first Cepheid in the Andromeda nebula, Hubble soon found more. With the aid of these wonderfully useful stars, he calculated that the distance of the nebula must be about one million light years, a figure that put the nebula far outside the boundaries of even the largest estimates of the size of our own galaxy. The nebula was, it seemed, not a mass of gas or a small star cluster within our own galaxy, but an independent stellar system—an island universe—in its own right.

With this new evidence, Hubble brought the long-running debate on the nature of the spirals to a quick end. The answer, almost all astronomers agreed, was that the spirals are indeed distant galaxies (although Hubble would never use this term, always preferring "extragalactic nebulae"). Evidence had pointed in this direction even before Hubble detected the Cepheids, but in the eyes of his colleagues, Hubble had provided the data that clinched matters.

The Hooker telescope had played a crucial part in Hubble's investigations, but his discovery was not just the consequence of a state-of-the-art telescope. He was not a particularly deft observer, but here, as he would often do, he displayed great insight in identifying an important problem. Also key was Hubble's drive and his carefully thought-out scrutiny of the Andromeda nebula. His discovery of the Cepheids was serendipitous, like many other pivotal discoveries in modern astronomy, but it was certainly not a fluke. Hubble's success was the result of a meticulously planned research program that also settled the question he set out to answer.

Hubble would be centrally involved too in the next big advance in the understanding of galaxies. For this part of our story we need to go back a decade or so to make the seemingly unlikely connection between studies of the so-called canals on Mars and the large-scale nature of the universe.

In 1912, Vesto Melvin Slipher, an assistant in Percival Lowell's Observatory in Arizona, made one of the most momentous finds in the history of astronomy, and it seemingly had nothing to do with Mars. The observatory's basement housed an enormous collection of photographic plates of planets, stars, and what are now known to be galaxies that the Lowell telescopes had taken over several decades. Many are spectrograms—images that show the rainbow-like fingerprints of these objects that contain information about their compositions and motions. Four of these spectrograms, the first of which was taken in 1912, are of the Great Nebula in Andromeda, as it was called then, and all of them rank among the most important photographs in all of 20th-century astronomy.

RED SHIFT

The key to understanding how we know we are living in an expanding universe is called the Doppler Effect. Named for one of the first people to study it as an acoustic phenomenon, it applies to any radiation that travels at a finite speed. This includes sound and light. The basic idea, which also stands at the heart of Einstein's theory of relativity, is that if a source of radiant energy—whether a train whistle, a car horn, or a star—is moving with respect to an observer, the energy it emits (in the form of light or sound) will change in proportion to the speed and the direction the source is moving. If an ambulance is rushing toward you with its siren wailing, the frequency (or musical pitch) of the siren will appear to rise. The faster the ambulance is moving toward you, the greater will be the increase. As the ambulance passes you, its velocity reverses. Because it is now moving away from you, the pitch of its siren will now appear to lower, again by a degree proportional to the ambulance's speed.

Because the effect is proportional, it is far less for light, which travels much faster. An ambulance moving toward you still looks like an ambulance. Say it is moving toward you at 100 kilometers per hour, which is about one-twelfth the speed of sound (1,250

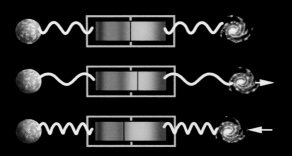

kilometers per hour). That is a noticeable shift. But the speed of light (300,000 kilometers per second) is so huge that the speed of the ambulance is miniscule by comparison. Nevertheless there is an extremely sensitive means to detect motion from light. Instead of using color (or pitch), physicists and astronomers use the positions of patterns of thin lines that appear in the spectra of almost all celestial objects, such as stars and galaxies, and arise from the presence in these objects of specific elements, such as iron, hydrogen, and oxygen. These lines are either bright or dark depending upon how they are produced and the structure of the source, and their positions in the spectrum are very precisely known from laboratory calibrations. If the same patterns of lines are found in the celestial sources, any deviation from the laboratory position will be proportional to the speed with which the celestial object is traveling toward or away from the earthly observer. If the object is receding, its "pitch" will be lower, as with the ambulance, and its spectral lines will be shifted to the red end of the spectrum. An object approaching Earth will have a higher than normal "pitch" and its spectral lines will be shifted to the blue.
—*David DeVorkin*

The direction and degree of shift in the lines of a spectrum can show that a star or galaxy is approaching (blueshift) or receding (redshift) and its velocity.

In 1903, the Lowell Observatory had purchased a state-of-the-art spectrograph to be used with the Lowell telescope's 24-inch refractor lens, one of the best of its kind at the time. By 1912, Slipher had thoroughly mastered the combination of the spectrograph and the refractor and had embarked on various research programs with them, including one on the nature of the spectra of spiral nebulae. Slipher was pursuing this research because Percival Lowell was fascinated by the evolution of the solar system—particularly the development of Mars. Lowell was convinced that the red planet's surface was crisscrossed by a remarkable system of canals, and that the purpose of these canals was to bring water from the planet's poles to its arid tropical and equatorial regions. For Lowell, the canals were signs visible even from the Earth that the planet was inhabited by intelligent beings.

Like the majority of astronomers at the time, Lowell also reckoned that the solar system had been born out of a swirling cloud of gas and that at one stage it had probably resembled the Andromeda nebula, so he directed Slipher to investigate the nebula spectroscopically. In his efforts to capture data on the nebula's spectra, Slipher believed he was probing the early history of the solar system. But what he found came to be widely interpreted as the critical evidence for a radical and completely unexpected view of the universe.

Slipher had, in fact, measured the "radial velocity" of a spiral nebula for the first time. The nebula was in motion with respect to Slipher, as the positions of the dark or bright lines in its spectrum had different values than they would have had in the absence of relative motion. The amount of the shift, called the Doppler shift, revealed the line-of-sight, or radial, velocity of the source. If a light source, such as the Andromeda nebula, is moving toward the observer, the spectral lines are shifted toward the blue end of the spectrum, or blueshifted, and if the light source is moving away from the observer, the spectral lines are shifted toward the red, or redshifted. To his surprise, Slipher measured the nebula to be rushing toward the solar system at about 300 kilometers per second, a speed faster than even the fastest moving star within the known Milky Way galaxy.

128

At Arizona's Lowell Observatory in 1912, Vesto Melvin Slipher painstakingly collected spectra for a number of spiral nebulae. He discovered that most seemed to be moving at astonishing speeds.

In the wake of his measurement of the nebula's spectral shifts, Slipher scrutinized additional spirals. These observations were far from straightforward, requiring that the light from the faint spirals be collected over many hours before the spectral lines became visible. One particularly long and demanding observation took some 80 hours. But all this hard work paid off. As the evidence from the spirals slowly accumulated, Slipher noticed a trend: Except for the Andromeda nebula, the great majority of the spirals exhibited redshifts. That is, they appeared to be fleeing our sun, usually with speeds far exceeding the fastest moving stars in our galaxy. What did this mean?

Some astronomers in the early 1920s wondered if the redshifts of the spiral nebulae might be indicators of a so-called de Sitter effect, named after the Dutch astronomer Willem de Sitter. Using Albert Einstein's theories, de Sitter had predicted that the intrinsic properties of space and time cause clocks to appear to run more slowly the farther they are from the observer. Following this same logic, the atomic vibrations within a far-off galaxy appear to slow down, the frequency of light decreases, the wavelength of light thereby increases, and a redshift is observed.

Growing interest among astronomers in de Sitter's effect led a number of them during the 1920s to seek out a correlation between redshift and distance. However, none of these efforts had been particularly persuasive. The resulting plots of redshift and distance presented a more random pattern, or what statisticians call scatter diagrams, so they were certainly not clear indicators of a relation between redshift and distance.

Hubble was naturally attracted to the puzzle of the redshifts and started in on them in 1928. While he studied the photographic appearance of the galaxies (their relative sizes and brightnesses, for instance) in his research, Milton Humason, another astronomer at Mount Wilson (who had begun his career on the mountain as a mule driver), measured the spectroscopic redshifts. In Hubble's first paper on this topic in 1929, almost all of the redshifts came from Slipher's data from the Lowell Observatory. Although the relationship he found between his improved distances and Slipher's redshifts could be discerned, there was still a lot of scatter in the data. But two years later he and Humason presented redshifts for the most distant galaxies ever measured, including one that Hubble estimated to be around 100 million light years away, moving (if the redshift was interpreted as a Doppler shift) at a staggering 20,000 kilometers per second. Now Hubble and Humason's work was definitive. No one doubted a clear relation between redshifts and distance. Simply put, a galaxy twice as far away as another would have twice the redshift. What Einstein's equations had predicted, Hubble had now proved: The universe was not motionless, but in fact, expanding.

By 1931, some astronomers, most significantly the Belgian abbe Georges Lemaître, were willing to identify the redshifts as Doppler shifts and to marry the relation between redshifts and distances with models of the universe derived from Albert Einstein's theory of general relativity. When Slipher had aimed his spectrograph at the Andromeda nebula in 1912, astronomers still were far from sure that external galaxies existed. By the early 1930s, however, astronomers could agree that the universe contained a myriad of galaxies. Moreover, many of these same astronomers judged that the galaxies had themselves revealed that the universe is not static but is expanding. And so was fashioned one of the great discoveries of

20th-century science. It also immediately raised a series of new questions, including "Will the universe expand forever, or will it eventually reverse its expansion and start to contract?"

Determining how fast the universe was expanding became the central goal in answering these deepest of questions. The universe's rate of expansion also bore on the issue of whether it is possible to identify the start of the expansion with the origin, maybe even the creation, of the universe. Theoretical astronomers and cosmologists tackled the creation question. Observers, of whom by far the most influential was Hubble, tackled the question of the rate of expansion. Indeed, after the publication of his first paper on the redshift-distance relation in 1929, his name became synonymous with the mathematical constant linking velocity and distance that determined the observed rate of expansion—the "Hubble constant." There was, however, a major difficulty. The value Hubble had derived for the constant seemed to imply that there was a time about two billion years ago when all the material in the universe had been packed together much more closely than it is now. From here it was a short step to talk about the "age" of the universe. The difficulty was that some estimates of the ages of stars in our Milky Way galaxy implied that the universe was much younger than the stars, clearly absurd! What had gone wrong?

Some cosmologists in the 1930s sought to avoid identifying the start of the expansion with the beginning of the universe. The brilliant English theoretical astronomer A. S. Eddington contended that, "As a scientist I simply do not believe that the present order of things started off with a bang; unscientifically I feel equally unwilling to accept the implied discontinuity in the divine nature." Eddington instead preferred a model of the universe in which it does not begin from a tiny, incredibly compact state (what Georges Lemaître referred to as the primeval atom). Rather, in this model, the expansion very slowly builds from a universe that already fills a vast volume and is initially at rest.

Hubble himself was always extremely suspicious of theories. His self-proclaimed duty was to observe more and more galaxies, firming up the empirical relationship that he had first found. In fact, Hubble was always careful in print to avoid definitely interpreting the redshifts as Doppler shifts. But, at the time of his death in 1953, almost all astronomers judged the redshifts to be Doppler shifts. The debate over whether the universe is expanding had effectively ended. By this time, the central question was whether or not the universe had begun in a "big bang" (as Lemaître had first argued in 1931) or whether it is in a steady state, in which the universe is effectively changeless because, as it expands, new matter is continually being created.

In the 1950s, more refined measurements were made of the distances of distant galaxies. There were problems with the Cepheid brightness calibrations, and, as it turned out, Hubble had mistaken so-called H II regions—huge superluminous clouds of bright ionized gas—for a galaxy's brightest stars. Because the clouds are intrinsically more luminous than even the brightest stars, it meant that the corresponding distances were underestimated. The result of these new measurements was that the difficulties with the timescale problem and the embarrassment of a universe younger than its oldest stars dissolved.

Astronomers still hotly debate the actual value of the Hubble constant. Widely varying results are constantly being published, though the variations have been narrowing over recent years. We turn now to one important reason why.

The value of the Hubble constant was very much a central issue in astronomy in the 1960s and 1970s when serious planning got underway for what astronomers would come to know in time as the Hubble Space Telescope (or HST for short), named after Edwin Hubble. By that time, Hubble's name symbolized the telescope's chief scientific goals, which centered on observing the most distant objects in the universe with unprecedented clarity. The HST traveled a very long and at times highly convoluted path from plans on a drawing board to the reality of an orbiting observatory hundreds of miles above the obscuring layers of the Earth's atmosphere. A good portion of my life as a historian of science is bound up with this telescope. I may well be regarded as one of the first "combat historians" in the history of modern astronomy. Much of what I say from this point on is based upon experience with the actual actors themselves, always a risky business for a historian.

131

The Hubble telescope began life in 1946 as the gleam in the eye of Lyman Spitzer, Jr., then a professor of astronomy at Yale. Spitzer soon after moved to Princeton University where he had trained as a graduate student. One of a generation of U.S. scientists who learned the fundamentals of science politics during World War II, Spitzer had a courteous and modest manner that belied a great determination to further those scientific projects in which he believed, and for such projects he was prepared to fight tenaciously. Spitzer, along with many other astronomers, fought extremely hard to

Lyman Spitzer, Jr., fought tenaciously for decades to realize his dream of placing a large telescope in orbit, where it could detect a wider range of wavelengths than a similar telescope on the ground.

bring the HST into being over a number of decades. At times the project was close to being cancelled. It survived, but on the road to approval from the White House and Congress it was scaled back in size from the original plans. At first its main mirror was to be 120 inches in diameter, but the telescope as built had a primary of 96 inches in diameter. What was then called the Large Space Telescope was scaled back to become the Space Telescope, though more than one astronomer had quietly regarded the first name, LST, as recognition of the driving force behind the instrument—the Lyman Spitzer Telescope.

One of the HST's great attractions to astronomers such as Spitzer was that, by getting

THE MODERN UNIVERSE EMERGES

Serviced periodically in space by Shuttle astronauts, the Hubble Space Telescope
continues to produce astounding images of the universe. A flaw in the shape of the
main mirror initially hampered the telescope's view, but a repair mission corrected
the Hubble's sight. Ironically, its backup mirror (above), now on display at the
National Air and Space Museum, is utterly flawless. Never aluminized, its clear
glass surface reveals the mirror's lightweight honeycomb structure.

FROM HUBBLE TO HUBBLE

above the atmosphere, it would be able to detect a much wider range of wavelengths than an equivalent telescope on the ground. Thus, while the 100-inch Hooker telescope at Mount Wilson was designed solely for optical wavelengths, the HST was fashioned as an optical, ultraviolet, and infrared telescope.

The HST is also a striking and hugely impressive piece of technology. In the 1980s, I followed the development of the telescope very closely as I researched its history. In 1985, I saw the telescope in a gigantic "clean room" at the Lockheed Missiles and Space Company in Sunnyvale, California. Held erect in a special mount, it was covered in silver-colored, multilayered insulation for the alternating swings from broiling heat to frigid cold as it slipped between the daytime and nighttime parts of its orbit. It seemed far bigger and more imposing than the 40-foot by 15-foot dimensions of its main structure would suggest. I was also struck by the HST as a marker of just how far telescope making had advanced during the 20th century. The Hooker telescope had first been turned to the heavens in 1917. The HST's primary mirror is slightly smaller than the 100 inches of the Hooker telescope and so is quite small by the standards of the biggest contemporary ground-based telescopes. However, in the intervening 70 years between the completions of these two instruments, a remarkable series of technological advances, combined with the willingness of NASA and the European Space Agency to spend billions of dollars on the project, have led to a vastly more powerful scientific tool. The telescope is entirely automated and is commanded from the ground, with astronauts periodically visiting it in orbit to make repairs and upgrades.

What I did not know at the time I first saw the telescope, and only one or two people suspected it, was that the HST's primary mirror had a flaw. It was a shade too flat at its edges by an amount less than a fraction of the width of a human hair. This nevertheless was a big error in the world of precision optics, and for its first three years in orbit following its launch in 1990, the HST was not nearly such a powerful telescope as its advocates had hoped. But a brilliantly successful repair mission by Shuttle astronauts in 1993 restored the telescope to its full capabilities, and some of the recovered pieces of the original telescope are now on exhibit at the Smithsonian.

If there was one scientific problem that became associated with the

For ten days in 1995, the Hubble Space Telescope focused on a tiny, relatively empty patch of sky and recorded every galaxy in its narrow line of sight. The resulting Deep Field image contains more than 1,500 galaxies—a "core sample" of the universe all the way out to its farthest observable limits.

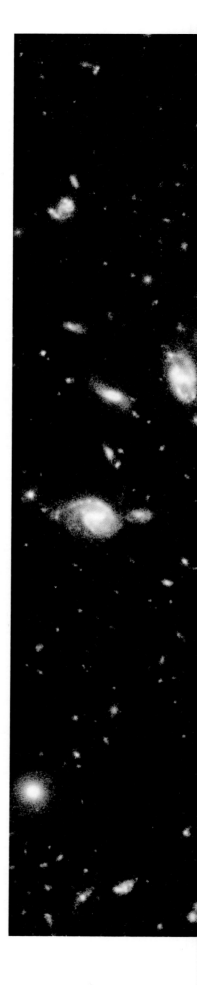

FROM HUBBLE TO HUBBLE

HST before it was launched, it was that of the determination of the distance scale of the universe and the determination of the Hubble constant. In 1977, when NASA issued a call for astronomers to become involved in the development of the HST, the determination of the Hubble constant was identified as the number one scientific problem. By 1984, this problem had become a "Key Project" for the HST. The goal was to come up with a value of the Hubble constant that was accurate to within 10 percent. Wendy L. Freedman of the Carnegie Observatories and her Key Project team painstakingly arrived at the value. The expansion of the universe, the "Hubble Flow," as it is called, increases by 70 kilometers per second for each mega-parsec increase in relative distance. A mega-parsec is equivalent to some 18,600,000,000,000,000,000 miles, a rather messy number. This implies that the universe is about 13 billion years old. However, while the current estimates of the Hubble constant now vary less than they did a decade or so ago, the problem has still not been solved in the sense of astronomers agreeing on a single value. Perhaps such a value will have to await future missions that will follow the HST into space.

In late 1995 the HST spent ten days pointed toward a patch of the sky about a tenth of the diameter of the full moon, near the handle of the Big Dipper. The aim was to provide a "deep core sample of the universe" by securing an image of objects as faint as possible with the telescope. The result was an image of the "Deep Field"—more than 1,500 galaxies at the very farthest limits of the observable universe. In Edwin Hubble's most famous book, *The Realm of the Nebulae*, published in 1936, he wrote about great telescopes and their role in expanding our knowledge of the universe by pushing deeper into new regions. For the HST, the Hubble Deep Field is perhaps the most spectacular example of exactly this, and is perhaps the telescope's most scientifically significant image to date. What one U.S. Senator dubbed as a "technoturkey" after the flawed mirror was first discovered, has proven to be the most famous and in some ways the most productive telescope ever built, one that has helped reshape and extend our views of the physical universe.

For a historian of astronomy, the launch of the HST stands as a watershed experience. For the first time I found myself in the position of seeing history made before my eyes. Surely the HST project was far too large for any one player to comprehend firsthand. But early on I gained access to masses of private and public papers that revealed the complex interrelationships of the players and the pressures that drove them to do what they did. Many of the principal figures in the history freely subjected themselves to highly structured formal tape-recorded interview sessions, knowing that I and other colleagues would transcribe and edit these into a permanent record. These interviews underline that history is a dynamic process, our view of which can change with time, distance, and perspective. Much like the universe itself.

As we look deeper into the universe, we also peer farther back in time. Sunlight is eight minutes old by the time it reaches our eyes. Starlight may be hundreds to thousands of years old, and the light from galaxies may be millions or even billions of years old.

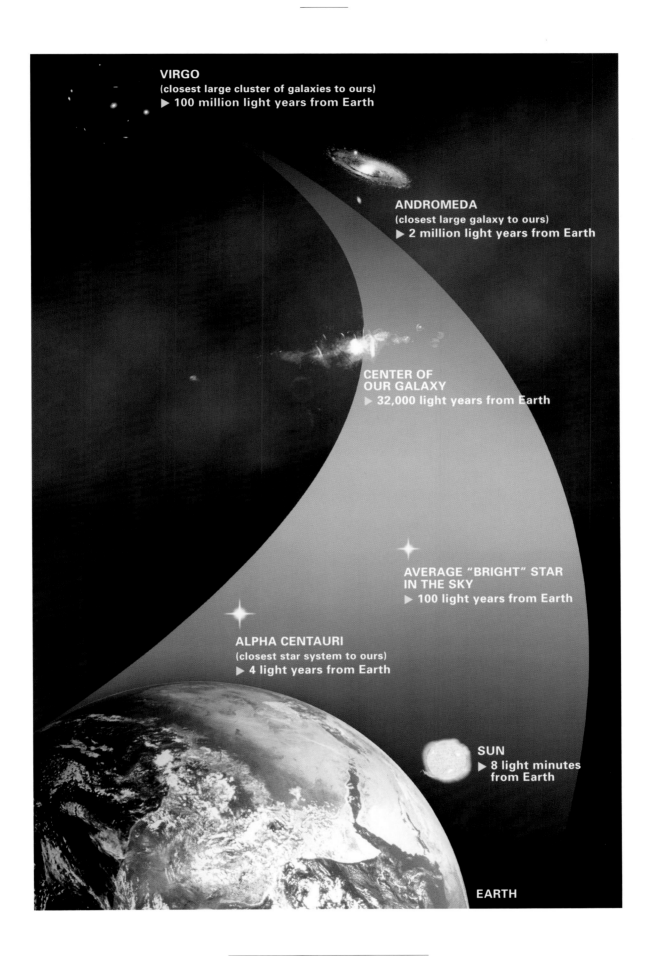

VIRGO
(closest large cluster of galaxies to ours)
▶ 100 million light years from Earth

ANDROMEDA
(closest large galaxy to ours)
▶ 2 million light years from Earth

**CENTER OF
OUR GALAXY**
▶ 32,000 light years from Earth

**AVERAGE "BRIGHT" STAR
IN THE SKY**
▶ 100 light years from Earth

ALPHA CENTAURI
(closest star system to ours)
▶ 4 light years from Earth

SUN
▶ 8 light minutes
from Earth

EARTH

137

WHAT IS THE
UNIVERSE MADE OF?

David DeVorkin

In the 1920s and 1930s, as observers and theorists mapped out a new expanding universe of galaxies, another revolution was taking place in astronomy. It was a revolution in our understanding of what the universe is made of—a revolution every bit as important as the expanding universe was, for both led to a single, consistent picture of how the universe evolved since the big bang.

What is the universe made of? Are the sun and stars and nebulae composed of the same stuff that makes up the Earth and the things on its surface, including us? Are the substances we are familiar with—such as water, alcohol, dirt, air, metal, and plastic—to be found on other worlds? If we are really a part of this universe, one would expect so, at least as far as naturally

In this view of the Sagittarius star cloud toward the heart of the Milky Way galaxy, the Hubble Space Telescope captured a glittering field of stars. Light from individual stars reveals their chemical makeup and provides clues to the composition of the universe.

occurring substances are concerned.

Today, everyone knows the universe is made mainly of hydrogen. It is found in water, in all organic materials, and in the stars. In high school, I remember doing a little demonstration separating hydrogen and then igniting it with a little "whoop." On science day in Santa Monica, I even spent some hours in a large display window case in our local J.C. Penney's store making hydrogen "whoop" for shoppers. I was rather surprised, then, when I got to college, to read in a textbook that not so long ago, the sun, stars, and most of the universe were thought to be composed of just about anything but hydrogen.

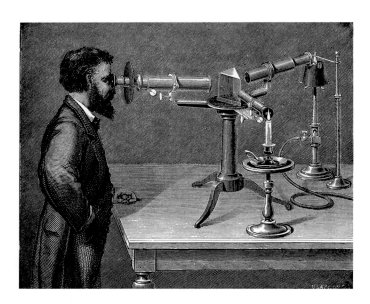

140

Using spectroscopes, scientists in the 1800s discovered unique patterns of lines in the spectra of chemical elements that could be used to identify those substances.

In fact, it was only in college that I realized that astronomy is really based upon physics, and that it was through physical arguments that the true composition of the universe was ever figured out. I recall being puzzled at first by all the physics courses I had to take, when what I wanted to do was look through telescopes. But the hydrogen question, among others, convinced me that looking through telescopes was only a beginning.

Only a few centuries ago, few if any people asked what the universe was made of. From antiquity, the celestial vault was made up of materials and substances either not found in this mortal world, or found in a state unobtainable on Earth. Kepler, Galileo, and Newton helped us think of the universe as a physical place—a realm subject to the same laws found here on Earth. But even in the 19th century, philosophers and theologians counseled that the study of the chemical and physical constitution of the heavens and the things in it was beyond human knowledge. Some even taught that it was not our mission to attempt to determine the physical substance of the universe—it was not a proper goal of science. It did not matter that Newton and his followers had extended the laws of physics to the stars, showing how they obeyed the laws of universal gravitation. Newtonianism consisted of the purely mechanical—a clockwork universe.

All of that changed in the 1860s when Gustav Kirchhoff and Robert Bunsen convinced everyone that each element produces a unique spectrum that serves as a sort of fingerprint for that element, no matter where it resides. Therefore the spectroscopic appearance of the sun and stars revealed the chemical elements in their atmospheres. "If we were to go to the sun, and to bring some portions of it and analyze them in our laboratories, we could not examine them more accurately than we can by this new mode of spectrum analysis," Warren de la Rue proclaimed as he spoke of Kirchhoff's work in 1861.

Just what is spectrum analysis? It all starts with the rainbow of colors contained in white light, which is revealed to us when that light passes through a prism. In nature, raindrops act as a prism, producing a rainbow when sunlight passes through them. Similarly, glass prisms spread white light out into its spectral colors. The presence, absence, relative intensities, and appearance of these colors in the spectrum of an object emitting light are all clues to the nature of that object. Using a spectroscope, consisting of a tiny mechanically operated slit, a collimating lens (which produces parallel beams of light), a prism, and a small telescope, Kirchhoff taught that a hot, solid body glowing enough to produce a spectrum will create a continuous, changing ribbon of color, like a rainbow. He also showed that a gas under low pressure, when viewed with the same instrument, produces a unique set of colored lines. The lines, he knew, were simply multiple images of the entrance slit to the spectroscope. When he viewed the same glowing gas while it was in front of a hotter solid, he found that the lines that had appeared bright against a dark background in the former spectrum now appeared dark in front of a bright continuum of color.

A spectrum can appear as a continuous wash of light, or contain patterns of bright or dark lines—"fingerprints" of particular elements present in the light source.

141

My first encounter with Kirchhoff's teachings came when, as an undergraduate student in an astronomy course, I was required to take part in an experiment illustrating this phenomenon. I remember how uncomfortable it was in the dark, little room where the experiment was conducted. Its three hot gas burners had been blasting away for hours as student after student entered the room to make observations. When I entered the chamber I found a strange instrument in front of me. It held the three burners in a line, like the smokestacks of an old ocean liner. At the far end of the device, in line with the burners, was a large incandescent light connected to a big dimmer dial, and, in front of the burners, was a small spectroscope on a swinging arm. The arm let the spectroscope swing into the line between the light and the three smokestack burners, where it could then be turned 90 degrees to allow the experimenter to look only at the burners from the side.

I first turned on the light by turning the dimmer dial. As the light brightened, I viewed it through the spectroscope and saw a continuous band of colors. When the dial was at a low setting, the spectrum was brighter in the red than in the blue. As I cranked it up to full intensity, yellow dominated over the red, and the blue remained faint. With the light on full, I was instructed to pick up a vial of sea salt crystals and shake some of it onto the mesh above the burners. Immediately a bright yellow light flared up over the three burners. I quickly rotated the spectroscope to the side and looked at the colored flame. Two bright yellow lines were all the spectral lines I saw. I then rotated the spectroscope from the side to the front, in line with the burners and the incandescent light. I peered through the spectroscope as I turned it, and saw the bright yellow lines turn dark as the continuous spectrum from the lamp came into view. Aha! The dark double line and the bright yellow double line were one and the same. Both were coming from the same substance—the salt in the flame.

Wiping my face, for it was incredibly hot in that place, I continued to look at the dark line spectrum and, when instructed to do so, turned down the dimmer without moving the spectroscope. The dark lines and the continuous spectrum disappeared, and I was left with the double yellow bright lines again.

This experience convinced me that the bright line spectrum of the salt—actually the sodium in the salt—appeared bright when seen alone, and appeared dark but in the same place when a brighter incandescent source was placed behind it. Kirchhoff taught that this is what stars do, and that by matching bright lines in laboratory spectra of common elements such as iron, carbon, nitrogen, and sodium to the dark lines found in the spectra of the sun and stars, one can deduce what substances are present in those celestial bodies.

In the 1860s, Kirchhoff managed to identify some 16 Earthly elements among the 600 lines he recorded visually in the solar spectrum; by 1891, that number had increased to 36. Science generally agreed with H. A. Rowland at Johns Hopkins, who stated in 1891 that "were the whole Earth heated to the temperature of the sun, its spectrum would probably resemble that of the sun very closely."

Astronomers who followed Kirchhoff's lead in Europe and America were less interested in the fine details of spectra. Instead, they gathered as many spectra as possible to see how they differed from star to star and, in the finest tradition of the scientific collector, carefully arranged their spectra into groups and series, creating systems of classification. Almost immediately there seemed to be a correlation between color and spectrum. Three great classes eventually emerged: the blue (or white stars), the yellow (or solar-type stars), and the red. By the end of the century, a vast photographic spectral classification program at the Harvard College Observatory was showing that hundreds of thousands of stars could be fit onto a systematic scheme of classification based upon the appearance of their spectral lines.

But what did these different types of spectra reveal about the stars, and why did stars have different spectra? Why did the blue and white stars show mainly lines due to hydrogen and little else, whereas the yellow stars (the sun among them) showed hundreds of iron lines, as well as calcium and some hydrogen. Did this mean that there were hydrogen stars, calcium stars, and iron stars? A few astronomers speculated that temperature was at play, but no one knew why this would be so. Nor did more than a very few test Kirchhoff's dictum of "one element, one spectrum," which was, after all, the whole key to the spec-

Distilling a spectrum from the feeble light of a dim star using photographic plates was no easy task. The huge 36-inch refractor at California's Lick Observatory, simulated in *Explore the Universe* as a backdrop to its famous Brashear spectrograph on display, provided the power to faithfully record stellar spectra.

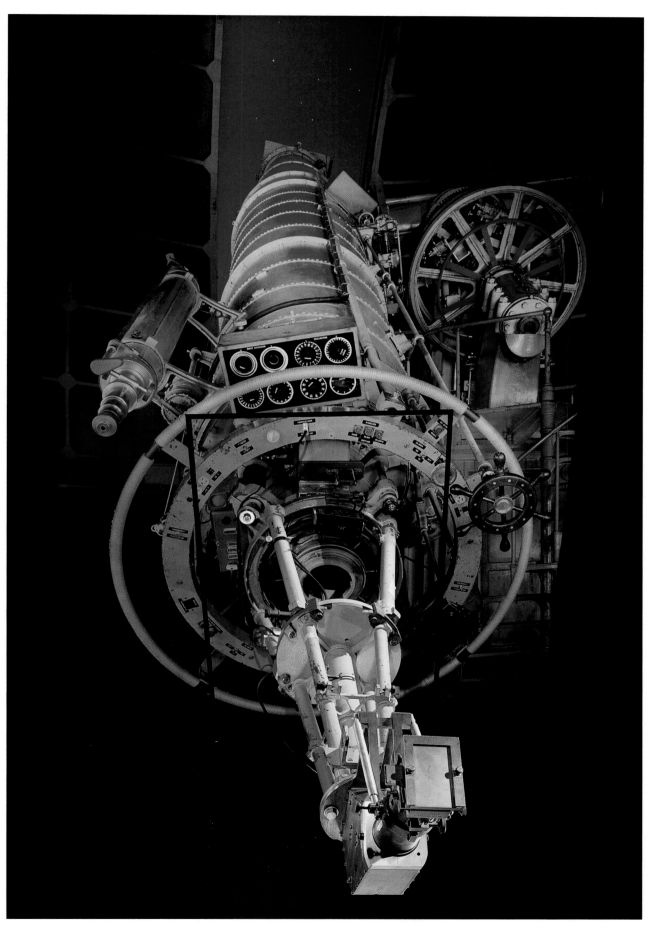

WHAT IS THE UNIVERSE MADE OF?

Spectrographs combine the elements of a spectroscope and a camera. They use prisms or gratings to spread out the light from the sun or a planet, star, nebula, or galaxy into a spectrum and then they record an image of that spectrum photographically on film, a glass plate, or now a digital detector called a CCD. The photographic images from a spectrograph, spectrograms, can be photographic records or electronic files.

Using the laws of modern physics, spectroscopic data can be interpreted to reveal the temperature, pressure, density, composition, and motion of celestial objects. Data from spectrographs have revealed that the universe is not static, but expanding, and that the chief visible component of the universe is hydrogen.

All spectographs have some form of entrance aperture, usually in the form of a rectangular slit, to let the radiant energy into the system. They all have a means of breaking down the radiant energy into its components, and they all have a means of recording the patterns of energy that are produced.

Each of the seven astronomical spectrographs displayed in *Explore the Universe* is famous for determining radial velocities, or motions in the line of sight, as well as generalized spectral characteristics. Some were optimized to study stars or galaxies, and others were designed to study the character of the cosmic microwave background radiation and even to search for evidence of black holes at the centers of galaxies.

One of the most famous spectrographs on display is the Prime Focus Spectrograph of the 200-inch Hale telescope from the observatory on Mount Palomar in Southern California. When this spectrograph was put into operation in 1951 on the Hale telescope, it created an unbeatable combination for the study of cosmologically interesting questions because, together, the two devices could image the spectra of the faintest known galaxies. Using the spectrograms the Prime Focus Spectrograph produced, astronomers measured the expansion rate of the universe with a precision never before possible.

The spectrograph had to be handled in total darkness and cold, so it was sturdy, smooth-edged, and massive enough to respond slowly to temperature changes, like the presence of body heat. It was also very compact and could fit into the confined space of the observing cage at the top of the telescope tube. In practice, the spectrograph sat flatly on a column in the observer's cage. Light from the Palomar mirror would enter from below and strike a small polished slit on the bottom of the black metal housing. Some of the light would reflect off the jaws of the slit and be imaged by an eyepiece that the astronomer would use to make sure the instrument was properly lined up. The light that passed through the slit was then reflected onto a concave mirror, which collimated the beam and sent it to a diffraction grating, which in turn acted like a prism to break up the light into a spectrum. That spectrum was then photographed by a very fast "Schmidt" camera loaded with tiny glass plates of photographic emulsion.

It took many hours, sometimes many nights, to collect enough light from an extremely faint galaxy to record its spectrum.

To record the spectrum of a galaxy, the light collected on the entire surface of the 200-inch mirror was concentrated onto a tiny photographic plate inside the camera that was less than an inch across. The resulting spectrum was even smaller—a third of that width. Lenses made of diamond and sapphire were added to the cameras to keep the spectrum in focus across the entire field.

This spectrograph played a critical role in deciphering a wholly new class of extragalactic object. In the 1950s, as radio telescopes grew more sensitive and capable of pinpointing discrete sources of radio energy, dozens and then hundreds of sources remained unidentified by optical means. The Palomar telescope was called to action, and, by the 1960s, astronomers such as Maarten Schmidt and Jesse Greenstein were finding that the optical smudges that seemed to be linked to the radio noise had radial velocities, or redshifts, far greater than any normal galaxies. Schmidt and his colleagues found what are now called "quasi-stellar radio sources," "quasars," or simply "QSOs." Now known to be an extremely energetic (or luminous) form of galaxy, QSOs have greatly extended our baseline for determining the character of the Hubble Constant.

—David DeVorkin

trum analysis of the sun and stars.

There were detractors to Kirchhoff's dictum who believed that they could make the spectrum of a substance change drastically by applying enough temperature. In the early 1870s Norman Lockyer, the founder-editor of *Nature* magazine, was certain that he had found spectral lines common to all elements when they were heated enough to reveal the basic structure of matter. What these units were, however, was impossible to say, because there was then no accepted theory of the structure of the atomic elements, or even of what an "atom" was. By the 1890s, however, high-quality spectra produced by large telescopes and spectrographs at the Lick Observatory, as well as high-quality solar spectra taken during eclipses, demonstrated that Lockyer's fundamental units of matter did not exist. So the "one-element, one-spectrum" idea held firm.

Meanwhile, other people were convincing themselves that matter did have a basic structure—not as Lockyer had supposed, but a far more orderly one. The hydrogen spectrum was very neatly arranged, as was helium. Gradually, as physicists examined these spectral patterns, they found rules that could predict where families, or "series," of lines might appear. This sort of thinking, plus some new experimentation on how particle beams react with thin metal sheets, led to estimates of the sizes of atoms and the idea that atoms were "nuclear" in structure—with dense centers surrounded by diffuse halos. Finally, Niels Bohr's work showed that he could reproduce the spectrum of hydrogen by creating a "quantized" atomic model. His model looked like a little solar system, where particles called "electrons" spun around a dense nucleus of material. The electrons could occupy only very specific orbits, representing specific levels of energy in the atom, and they could only move from one level to another, either propelled by a discrete amount of energy from a photon, or releasing that same amount as a photon. Bohr realized that these electron jumps could cause both absorption (dark) line spectra and emission (bright) line spectra.

In any atom, the energy levels that are permitted to exist do not change, but the configuration of electrons can change due to the application or restriction of energy. Bohr and other physicists realized this. So did a young radical physics student in Calcutta. In 1983, I was invited to India to give a series of talks about this brilliant man, Megh Nad Saha, and how he linked the physicist's laboratory bench to the stars. That trip was a life-changing experience for me, as indelible as my sweaty encounter with Kirchhoff's laws. Whether I was in Pune, New Delhi, or Calcutta, my audiences were always enthusiastic because I was talking about an Indian hero. During that trip, I managed to avoid the snakes that dropped out of trees, and the taxis that dodged the cows that parked in the middle of the road at night, honking at them instead of using their lights because they believed honking used less fuel. But the poverty and the press of humanity were unavoidable, and made me wonder more than once about the priorities of physicists and astronomers in India and elsewhere.

I knew that Saha was a national hero not for his physics, but for his social conscience. He may have been the first person in the world to show, in a set of papers written in Calcutta between 1919 and 1920, how to estimate temperature and pressure from analyzing the spectra of stars, and even how to estimate their chemical compositions when looking at them collectively. But to the Indians he was best known as a populist parliamentarian, a radical

146

lower-caste Bengali who campaigned for cultural reform and for the domestic application of Western science to the needs of the nation. After India won its independence, he held national office until his death in 1956 at the age of 62.

Saha's work was breathtaking. What he started was extended by English and American astronomers and physicists. It was a revolution in miniature because it showed how the study of the atom was indelibly linked to the study of the stars. It was both a revolution in technique as much as a revolution in attitude, and forms the basis for the emergence of modern astrophysics as a distinct discipline. In 1924 E.A. Milne, a follower of Saha's work in England, looked back over the previous decade, noting the many correlations that had been found in stellar spectra, and the concern of many that these correlations lacked a rational basis:

> *It was known that some spectral lines could be produced, in the laboratory, only at high temperatures or under intense discharges, and that such lines were often only to be found in stars with high effective temperatures. But it was not known why this was so, or why the same line tended to disappear at still higher temperatures; and of quantitative explanation there was none. There appeared to be a definite relation between effective temperature and type of spectrum, but the connection was empirical. There was a gap in the logical argument.*

Known as a social reformer in his native India, Megh Nad Saha also showed how to estimate the temperature, pressure, and composition of stars by analyzing their spectra.

Saha closed this gap by linking Niels Bohr's theory of the atom to chemical ideas about chemical equilibrium, and then showed how the behavior of spectra in the venerable classification sequence created at Harvard was actually a sequence defined by temperature. As one moved through the spectral classes, from the coolest red stars to the hottest blue and white stars, spectral lines of the different elements would appear, grow stronger, and then weaken and vanish. This behavior, as Saha showed, was due first to the growth of the fractional number of electrons in the right energy states to absorb the particular energies that produced the visible spectral lines. As temperature increased through the classes, there came a point of greatest line intensity when the fractional population was at a maximum. The strength of the line would diminish with continued elevated temperature because the specific energy states would be depopulated by the loss of electrons. Saha was the first to break through, and so brought the universe down to Earth and accessible to quantitative methods of analysis. There was still an enormous gap between the laboratory and the stars, but it was closing.

Saha's work had a profound influence in the West among physicists and astronomers. It even caused some astronomers, such as Henry Norris Russell of Princeton, to change their career directions and research interests. Russell saw the power in Saha's analysis, and believed that he had opened a whole new realm to analytical investigation, which he indeed had. Within the decade, for instance, followers of Saha's theory completely revolutionized our understanding of the composition of the universe. Although Saha's theory was far from perfect, his physical intuition was impressive, and everyone agreed that he had opened the right door. While British theorists concentrated on rederiving Saha's equations using more solid

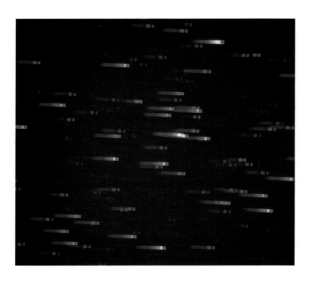

mathematics, in the United States, Russell and others went through the vast stores of spectroscopic information at observatories such as Mount Wilson and Harvard. Russell was soon able to verify several spectroscopic predictions Saha had made, but just as quickly found problems—anomalies in the spectra that Saha's theory could not explain. Two of these anomalies were especially bothersome, one of which concerned hydrogen.

In the early 1920s, such astronomers as Anton Pannekoek in Holland, H. H. Plaskett in Canada, and Cecilia Payne in the United States all wondered about hydrogen. Payne, a student of some of the best theoretical minds in Britain, had moved to Harvard to apply her fine training to the vast stores of data there. After applying her highly refined version of Saha's theory to the mass of stellar spectra at hand, she was able to not only improve the temperature calibration for all stars, but also show what the relative abundances of the elements were in their atmospheres.

Payne used a technique suggested by Saha, one he called the "marginal appearance" of a line. Saha used his theory to calculate the temperatures at which various elements appeared and disappeared in the spectral sequence. He wanted to compare these points to actual observations, realizing that discordances would be a measure of relative abundance, but without sufficient data was never able to do so. Still, he published his plan, and Payne followed it up. She knew that if the observed appearance was lower, or the disappearance higher, the element was overabundant. This was the case for hydrogen. So much so, she found, that hydrogen was by far the most abundant element in the universe, far surpassing any other, with helium running a distant second.

After many months of toil in the plate vaults in the library and at her desk at the Harvard College Observatory, Payne wrote up her conclusions in late 1924 and gave them to her advisor, who sent them to his old teacher, Henry Norris Russell. Russell was fascinated and impressed by Payne's work. He had already recommended her for a fellowship to extend her stay in the United States, and he approved of most of her conclusions, mainly for the relative abundances of the heavier elements. Even though he advised her that reliable abundance determinations required more physics than was yet available, he was impressed with what she had been able to do with Saha's imperfect theory. In fact, most of her abundances

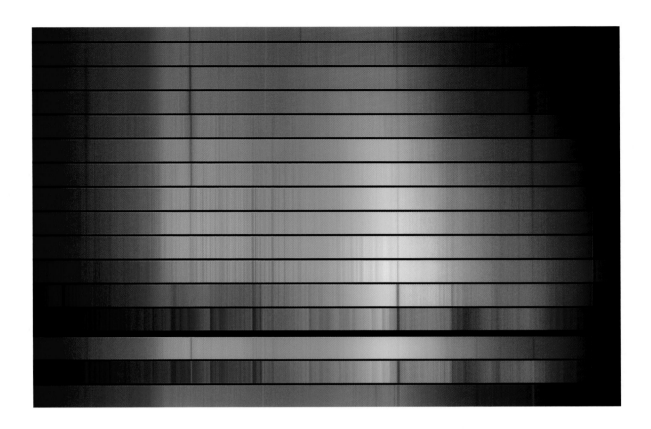

were not far from what he had suggested in previous work, based upon non-spectroscopic criteria. But there was one element where her technique had surely led her astray, which implied that something was still very wrong with the overall theory. "It is clearly impossible," Russell wrote Payne in January 1925, in a letter that has been preserved in the Princeton archives, "that hydrogen should be a million times more abundant than the metals." He was firm in this conviction because so much of his own work, and that of Arthur Stanley Eddington, the most influential astrophysical theorist of that day, required that hydrogen not be overly abundant. The arguments were legion, and Payne well knew them. So even though she was confirmed in her opinion of her own conclusions, she let them remain published in her thesis, but marked as illusory, in line with Russell's wishes.

Even though Russell was wrong, he was also right given the situation, and Payne was very wise to take his suggestion. Saha's technique, even in the very careful hands of Cecilia Payne, still rested on very incomplete theory. No one could yet describe in a completely sat-

By placing a thin prism in front of a telescope lens (left), Harvard astronomers photographed the spectra of thousands of stars and classified them into a sequence represented by the 16 spectra stacked above. The arrangement is based on variations in strengths of hydrogen lines (the vertical lines in the spectra above) and was shown in the 1920s to be a physical temperature sequence first by Megh Nad Saha, and then refined by Cecilia Payne.

isfactory way how many atoms of an element were required to create a visible spectral line. Still, Payne's work was exemplary; her thesis and political acumen won her a place on the Harvard Observatory staff. She went on to the most distinguished career a woman could hope for in those days of male dominance. Today she probably would have been chosen to direct the observatory.

Studying the spectra of stars, Cecilia Payne refined the method used for determining the temperature of stars and deduced the relative abundances of elements present in their atmospheres.

Russell spent the next four years trying to make hydrogen go away. He was not alone; most of the best theorists in Europe tried hard as well. Meanwhile, Payne continued to think about the problem, and Russell's students began finding evidence that pointed in her direction. And every time Russell attacked the problem, collaborating with astronomers at Mount Wilson and physical theorists on the East Coast, hydrogen came out very abundant. Finally, by late 1928 Russell reversed course and marshaled all the evidence at hand. Every argument that he had used in the past several years to make hydrogen go away he now used to show that hydrogen was king. "The conclusion from the 'face of the returns,'" he wrote in early 1929, after months of struggle he called an "orgy" of work, was that hydrogen was at least 80 times as abundant by weight as all the metals together.

Specialists closely following this chain of events also knew that Russell had gone as far as anyone could using traditional lines of argument. His patchwork of arguments, which he later referred to as a "tissue of approximations," was not as convincing as the application of the most modern forms of the new quantum theory, which appeared just as he published his paper. But they all confirmed his numbers. So what is now known as the "Russell mixture" of the elements in stellar atmospheres remained the accepted composition profile of the universe for many decades.

During the 1930s, astronomers and physicists also realized that their models of stars worked better if they had mainly hydrogen throughout their volumes. In other words, hydrogen dominated throughout the stars, not just in their atmospheres. And finally, in 1939, Rupert Wildt, a refugee from the Nazi terror, showed how a stellar atmosphere governed by Russell's mixture produced a spectrum that agreed with observations. Up to that point, although abundances were manifest by the strengths of the spectral lines, no one had been able to make the general appearance of the spectrum agree with theory. Wildt finally did it using quantum mechanics to show that a fair fraction of hydrogen atoms in a star's atmosphere had an extra electron weakly attached—so weakly, in fact, that it could be easily broken away by photons (particles of light) over a wide energy range. For this reason, hydrogen with an extra electron absorbs energy over a broad range of wavelengths, including the entire visible spectrum. This kind of continuous-spectrum absorption—or "continuum absorption," as spectroscopists call it—causes stars to appear solid, much like a cloud in the sky looks solid (though the mechanisms responsibile for these

illusions differ). What astronomers viewed with their eyes and in the spectra of stars could at last be reconciled. This new view of our material universe has not significantly changed. We do live in a universe made almost completely of hydrogen. Some form of hydrogen fusion dominates in most stars, and the best theories of the big bang argue that the only elements synthesized in the early moments of the universe were hydrogen and helium, with a whiff of deuterium. The universe expanded and cooled too fast to allow for the heavy metals to form. These heavier elements, the ones our bodies are made of, as well as those making up the Earth, came into existence only through sustained fusion processes in the centers of stars over millions of years, and in the brief but enormously violent explosions of stars called supernovae.

All of this was made possible by thinking of the sun, stars, and galaxies as physical systems, based upon physical law and described by the laws of physics. This new way of thinking, less than two centuries old, still only experienced its full flowering in the last 50 years, for only recently have astronomers learned to think first in terms of physical processes and physical questions when framing research projects. Their tendency to do this today is as profound as any revolution in astronomy has been. Asking "What is the universe made of?" has transformed the way we think about the universe.

At the Mount Wilson Observatory, physicist E.S. King built a laboratory to reproduce the spectral properties of solar and stellar atmospheres. His laboratory became a center for the study of stars and atoms.

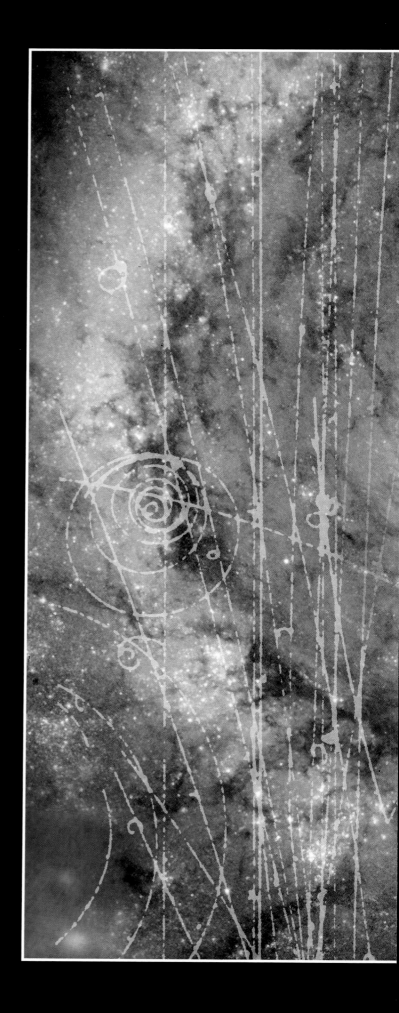

SECTION III

THE CURRENT UNIVERSE

Superimposed on an image of the Whirlpool
galaxy, particle tracks from a high-energy
accelerator symbolize the many ways we examine
physical processes in the universe today.

THE CURRENT UNIVERSE

David DeVorkin

Just as spectroscopy broadened our horizons in the early 20th century, the growth of electronic and radio technologies, coupled with access to space made possible by military technologies and priorities in the Cold War era, vastly expanded the spectrum of radiation available to astronomy. At the same time, Hubble's correlation of recessional velocities of galaxies with distance quickly constrained possible world models to those that described an "expanding universe." Albert Einstein had already set in motion alternative universes to Newton's that allowed for this expansion. But at first he did not think it was such a good idea.

Einstein explored the universe not with a telescope, but with his mind. During the first two decades of the 20th century he developed his theories of special and general relativity; the latter showed mathematically how gravity, space, and time are related. According to Einstein, gravity is the warping of space and time by matter, and the warping of space affects how matter and light move. His special theory included the famous equation showing that mass and energy are equivalent ($E=mc^2$) but his general theory also suggested that the universe might be expanding. It was not until the 1930s that experimental physicists were able to actually see the process of the conversion of light into matter (pair production). But well before then other predictions he had made had been verified by astronomical means.

Einstein predicted that strong gravitational fields—those surrounding stars like the sun—could measurably alter the direction of starlight passing in their vicinity. He also explained an age-old anomaly in Mercury's motion around the sun. The same warping of the space-time continuum that bent the light from stars also altered how Mercury moved around the sun. The discrepancy was neatly matched by a correction derived easily from Einstein's equations.

When Einstein developed his general theory prior to 1920, he assumed that the universe was static, as Newton had long taught. But his equations indicated otherwise—that there would be a general expansion of the space-time continuum. To rid his theory of this annoying detail, he added a "cosmological constant" to balance the force of gravity to keep the universe static. In several years, other theorists such as the Russian Alexander Friedmann reexamined Einstein's theory and warmed to the idea that the universe could actually be expanding. By then the unsettling spectroscopic observations of V. M. Slipher of the high velocities of galaxies had stimulated some speculation by a few European astronomers and mathematicians that expansion was possible. So Friedmann had good reason to at least suggest the possibility, if only to distinguish his work from Einstein's.

Only seven years after Friedmann announced his work, in 1929, Edwin Hubble announced his spectacular findings. In the wake of Hubble's announcement, Einstein has been often reported to quip that his ad hoc cosmological constant was the biggest blunder of his career. The irony of this, as we will learn in David Wilkinson's essay, is that Einstein may be right after all—qualitatively at least. Not only does the expansion rate of the universe seem to be accelerating with age, perhaps due to the presence of some form of repulsive

155

force described by Einstein's "cosmological constant," but a detailed analysis of the distribution of density in the early universe hints at it as well.

In fact, the very idea that the universe must be in general expansion raises the question of what the universe looked like in past time. The implication of expansion is that our universe began expanding billions of years ago in a cosmic explosion known as the "big bang." This idea was first suggested in 1927, and then more forcibly in the 1930s. The big bang set in motion the expansion of the universe and as it expanded it dropped in temperature and density, eventually giving rise to everything from galaxies to stars to planets to life.

The idea that the universe began with an explosion and has been evolving ever since was not greeted with unalloyed enthusiasm by everyone. In the 1940s astronomical theorist Fred Hoyle chided the idea, calling it "this 'big bang' idea." He proposed an alternative model,

Thanks to the work of Edwin Hubble and others, the universe model that emerged in the 1920s still stands: an expanding universe in which our Milky Way (seen here at center) is one of countless galaxies moving away from one another. The rate of expansion—the Hubble Constant—is still a topic of lively debate.

still apparently expanding, but continually replenishing itself and thus remaining static in density, existing in a "steady state." This was a far more appealing model to many people because one now did not have to deal with origins. To the dismay of those who supported the explosion theory, "big bang"—the name Hoyle coined to mock it—became a standard term for describing the event.

At the time that Hoyle and his colleagues proposed the steady state theory, physicists working in the Washington, D.C., area developed a theory about how chemical elements formed during what is now called the big bang, trying to account for the observed abundances of the chemical elements in the universe. What George Gamow, Ralph Alpher, and Robert Herman concluded was that some of those elements, along with faint traces of heat from the event, should still be detectable if the proper form of sensitive radio receiver were built. Gamow whimsically dubbed the substance from which the universe formed "Ylem" ("EYE-lem," an ancient term for "the primordial substance") and relabeled a Cointreau liqueur bottle in its honor. In a photo taken around this time, Gamow appears as a genie rising out of the bottle between Herman and Alpher. The original bottle appears in the exhibition. The essays by experimental physicist David Wilkinson and by Robert Wilson, co-discoverer of the primordial radiation, will elaborate on its detection in the 1960s, and how it has been studied in following years from observations by ground-based telescopes as well as those carried aloft in aircraft, balloons, and rockets.

The appearance of a "nova" or "new star" in the constellation of Cassiopeia in 1572 prompted Tycho Brahe to search out its distance. His precise observations convinced him that the apparition did not reside in the Earth's atmosphere, but was beyond the moon, or in space. Aristotle had taught, however, that the heavens were changeless, but Tycho showed that this was not the case. Since then, novae have been studied for a wide range of interests. We know now that they are not "new stars" but older, evolved, massive stars in the process of violent collapse and disruption. Two very different classes of novae exist, and there are numerous subclasses. Each arises from a range of disruptive mechanisms, but all of them yield critical information on such cosmological questions as the origin and rate of evolution of the chemical elements and mechanisms for determining the relative ages of stars and galaxies.

Supernovae are rare events in any single galaxy; one must wait for centuries and even millennia before a sufficient number of massive stars blow themselves up. Astrophysicist F. Richard Stephenson, accordingly, set himself the task of uncovering records of historical supernovae in the oldest records and hit paydirt in Chinese archives. Here he recounts his collaborative efforts with a wide range of scholars to unravel this astrophysically important historical record.

Margaret J. Geller, a major practitioner in the technique of three-dimensional mapping of the large structure of the universe, recounts how her fascination with mapping led her to map out bubbles and voids in the extragalactic universe, and how that cartographic vision has played out in more recent surveys. The structure she and her colleagues first detected has now been found in the earliest visible moments of the universe, when the cosmic microwave background radiation was mapped out in the 1960s through the present day. We will be given a detailed accounting of the detection of this early structure by Princeton experimental physicist David Wilkinson.

Up to now we have been dealing with a universe we can somehow "see" directly, through the radiation emitted by its many parts, from long wave radio energy to the highest-frequency gamma rays. Over the past few decades, however, evidence has been growing that there is a substantial part of the universe that does not radiate energy in known wavelengths so far uncovered by our technologies. Now it seems that the universe is actually dominated by this "dark" matter. Though the universe is still expanding, its primary constituents may not be, after all, the visible galaxies, but entities far larger and more massive than anything ever "seen."

157

Carnegie Institution astronomer Vera Rubin has been credited as the discoverer of this new component of the universe. She is, more probably, the first person to detect its presence in a way that could not be ignored by science; in other words, she proved that something is out there that we cannot yet see. She shares here her excitement at detecting unseen mass surrounding galaxies and the influence it had upon her life and science. J. Anthony Tyson of Lucent Technologies, one of the most energetic explorers of dark matter, is among those who are searching out new ways to use Einstein's predictions as a method to map out the large-scale structure of the dark universe. Here he describes this effort, leading to what the astronomical community has endorsed as a new "dark matter telescope" that may well become a new defining tool in refining and extending the observation of structure in the universe.

Science fact surpasses science fiction in this fanciful depiction of a drive-in movie.
By the 1950s, astronomers had confirmed that that the galaxies were indeed flying
away from us. Technological advances in tools and new information set the stage
for a multitude of fascinating discoveries in the latter part of the 20th century.

LEARNING FROM THE PAST: HISTORICAL SUPERNOVAE

F. Richard Stephenson

To the ancient Chinese, they were *kexing*, "guest stars." Appearing suddenly in the night sky, kexing burned brightly for several months. If they were close enough to Earth, they would be one of the brightest stars in the sky—but only for a short time. Scholars meticulously recorded their positions, and struggled to discern whether their appearance was an auspicious sign for the emperor—or a portent of doom. Then, as suddenly as they appeared, the kexing disappeared, fading to a dim spot in the sky. Sometimes they disappeared altogether.

Today, we know that what the Chinese observed as astronomical "guests" were supernovae, the catastrophic death and energy release of relatively massive stars. Outshining galaxies, they are visible over vast reaches of space, and tell us a great

The Chinese divided the sky into 28 sections called lunar lodges, some of which appear in this detail of a star map inscribed on a tomb ceiling around 25 B.C. Historical records of "guest stars" appearing in a lunar lodge prove invaluable in tracking down the remnants of supernovae, giant stars that self-destructed.

deal about the evolution of the universe, as well as the birth, death, and metamorphosis of the stars within it. At its maximum brightness, a typical supernova radiates as much energy as 1,000 million stars like our sun, and may continue to shine for many months. Because of their extreme brilliance, supernovae are often detected telescopically in distant galaxies-almost at the edge of the known universe.

At the core of the Crab nebula, the shreds of a giant star that exploded in 1054, lies a superdense, rapidly spinning pulsar (the left of the pair of stars near the center) only a few miles across.

Supernovae provide a valuable tool for studying the universe and its history. Supernova explosions are responsible for scattering heavy elements—produced in the interiors of stars—into space. These elements become part of the material that forms new planetary systems. As the cloud of dust and gas released in a supernova outburst gradually expands, it becomes a powerful source of electromagnetic radiation at a wide variety of wavelengths (notably radio waves and x-rays) for many thousands of years. Studies of this radiation provide astronomers with much information about the nature and evolution of the remnants of supernovae. In some supernova explosions, the original star is completely disrupted, leaving only an expanding shell of matter. In others, the collapsed stellar core forms a neutron star, which acts as a source of pulsed radiation—a pulsar.

Within our own galaxy—the Milky Way—a supernova has not occurred for almost 400 years. The last supernova was observed in 1604 by many European astronomers, notably Johannes Kepler, (it is now known as Kepler's supernova) about five years before Galileo first turned his telescope to the skies. Consequently, present-day astronomers must rely on naked-eye observations made hundreds of years ago for information on the outbursts of supernovae in the Milky Way.

In total, only five such events have been recorded since A.D. 1000. These occurred in the years 1006, 1054, 1181, 1572, and 1604. During the preceding millennium, perhaps another three were observed, in A.D. 185, 386, and 393. All these supernovae appeared as very bright stars, sometimes visible in daylight. Most were seen for many months and some for several years. Extensive radio and x-ray observations in recent years have identified more than 200 supernova remnants in our galaxy, and scientists have cataloged their properties. Unfortunately, in all but a few instances, the parent star can no longer be identified, so the age of most supernova remnants can be only roughly estimated—from their expansion rates, for example. However, when a star recorded in history as "new" can be shown to be the parent of a present-day supernova remnant, we know the precise amount of time that the remnant has been expanding. This provides important information on the dynamics and energetics of supernovae.

Both supernovae and their smaller counterparts, novae (estimated to have a brightness of one-millionth that of supernovae), were typically described as "guest stars" in Chinese. Today, when astronomers examine early records, we look for two indicators that a guest star was a supernova rather than merely a nova: proximity to the galactic equator and long duration of visibility (sometimes many months). By comparison, novae usually remain visible to the unaided eye for only a few weeks, and often appear far from the galactic equator. Nevertheless, it is important to critically examine any close agreement in position between a guest star and a known supernova remnant regardless of how long the star was visible. This has been one of my main areas of research for many years, and it is here that my own story begins.

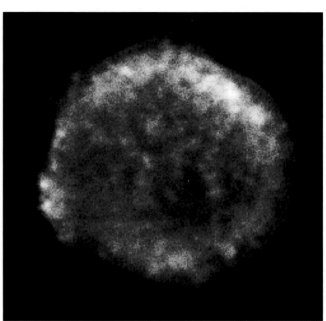

Danish astronomer Tycho Brahe witnessed the appearance of a supernova in 1572. The expanding shell of gas from the explosion is clearly evident in this x-ray image.

Somewhat rarely, a doctoral student has the good fortune to be assigned a research project that seems tailor-made—and, furthermore will keep him or her employed for the next few decades. I regard myself as extremely fortunate to have come into that privileged category. In the late 1960s, as a rather raw student in the newly formed Department of Geophysics and Planetary Physics at Newcastle University in the north of England, I undertook a project in a new research field that would later became known as "applied historical astronomy." My background was in physics, but when my supervisor— the late Professor S. Keith Runcorn, an eminent geophysicist—learned that I had an interest in astronomy, he suggested that it might be worth studying ancient eclipse observations, which can be helpful in determining the magnitude of the Earth's ocean tides in history. Tidal influences have been known to change the rate at which the Earth rotates, which in turn has changed the length of the day.

Because he was frequently away from campus, traveling especially to the United States, Professor Runcorn was never a thorough supervisor. He used to delight in telling the story of a former Cambridge academic who had the habit of briefly introducing a research project to a new Ph.D. student and then coming back three years later to remark: "That's not what

I told you to do!" As many of Runcorn's former research students will affirm, he himself came into much the same category.

Early in my studies, a senior member of staff at Newcastle University expressed his concern as to whether I "had enough material to get a Ph.D. out of it." However, today, more than 30 years later, I am still actively involved in the study of historical astronomical observations. Indeed, applied historical astronomy has become a full-time pursuit for me, and over the years I have published several books and numerous research papers on this fascinating interdisciplinary subject.

My initial research was in the study of ancient eclipse records, many of which are found in Oriental history. However, I soon became aware of the potential of the "new star" observations that I sometimes found in these same sources. Because Runcorn had given me very much a free hand, this aspect of my research began to take up a significant proportion of my time. In 1971, while still a graduate student, I made a detailed study of the Chinese and Japanese records of the new star of 1181. I was also able to propose a remnant of this star (called 3C 58 because it is the 58th object in the Third Cambridge Catalogue of Radio Sources), an identification that still finds general acceptance.

When I initially began exploring historical records of supernovae, my first task was to learn enough classical Chinese to cope with the necessary texts. This was the common written language of China, Korea, and Japan until relatively recent times, and it has the attraction that grammar is essentially nonexistent. Each individual character represents an idea and, in an astronomical record, one can usually (but by no means always) expect the same character to have the same meaning. In my first ventures in Chinese I was fortunate to encounter Archie Barnes at the School of Oriental Studies at nearby Durham University (where I have now been based for almost the last 20 years). He gave me his time unstintingly, and many an afternoon I would take the train from Newcastle to Durham and pore over Chinese texts with him. On other occasions I would telephone him and try to describe certain complex characters—no easy task. To his credit, Archie Barnes never gave up.

Surprisingly, I soon learned that a limited vocabulary was sometimes an asset. Because many astronomical reports were mixed with more mundane material, it was easy for the

162

This 12th-century text records the appearance of the supernova of 1181. A supernova visible within our own Milky Way galaxy is exceedingly rare—only five have appeared in the last thousand years.

competent sinologist to become sidetracked, but with my limited knowledge of Chinese (about 1,000 characters) I was able to focus readily on the items that concerned me most. There were some unexpected difficulties, however. Classical Chinese is completely unpunctuated and entries may run for several pages without interruption, so it is sometimes hard to determine where a particular entry ends and the next one begins. The similarity between certain characters can also lead to confusion—for instance, in determining the date of an observation. I have ample experience of both problems!

In addition to the necessary linguistic studies, I also had to familiarize myself with the various East Asian sources of astronomical records. Because of the early discovery of printing in China—around A.D. 700—it is rare to encounter manuscripts from this part of the world. To paraphrase Joseph Needham, most early Oriental records fall into one of two categories: printed or lost. Fortunately, high-quality editions of a wide range of East Asian histories are available in major libraries worldwide.

For China, from around 200 B.C. until A.D. 1368, the principal source of astronomical observations is the official dynastic histories. Most of these histories were written by teams of scholars not long after the fall of the dynasty to which they relate, and they usually include a detailed astronomical treatise, largely composed of the reports of the court astronomers who were based at the imperial observatory in the capital. Independent observational information sometimes appears in the imperial annals and occasionally even in the biographical sections of the dynastic histories. For the Ming dynasty (1368–1644), a major source is the *Ming Shilu* (*Annals of the Ming Dynasty*), an extensive series of annals covering the reign of each Ming emperor.

Korean history contains little in the way of original astronomical records until around A.D. 1000. However, for the next 400 years or so (until 1392), the *Koryo-sa* (*History of Koryo*) is an important source. This work is modeled on a typical Chinese dynastic history and contains a substantial astronomical treatise. For the Yi dynasty, which followed the Koryo period, the most extensive historical source is the *Yijo Sillok* (*Annals of the Yi Dynasty*), a major work following the pattern of the *Ming Shilu*.

In general, Japanese history is much less systematic than that of China or Korea. This is especially true of astronomical observations, many of which are scattered in such works as privately compiled histories, diaries of courtiers, and temple records. Fortunately, around 1935, Kanda Shigeru published a careful compilation of the celestial reports of his country through 1600, and, quite recently, Osaki Shoji continued this compilation through 1860.

Many historical texts have been lost as the result of wars, deliberate destruction (as authorized in 213 B.C. by Qin Shihuang, the first emperor of China, for instance), and careless storage. Around 1592 the preservation of the Korean *Yijo Sillok*, the historic record of the Yi dynasty, was in jeopardy, with only a single set of volumes surviving the Japanese invasion of Korea in that year. By that time, the records show, a complete set of *Sillok* had accidentally been destroyed by its keepers when they tried to smoke pigeons out of the building where the documents were housed!

Some secondary sources were available, but I quickly learned that there were hazards involved in using them. On one occasion I remember well, I discovered a report that seemed

On a February night in 1987, a gigantic flood of neutrinos passed through the Earth. Fifty trillion of these elementary particles went through each and every one of us. Only 19 were caught by special detectors—most in a huge pool of pure water near Kamioka, Japan. The detectors were completely automatic and untended at the moment. Minutes later, a blast of visible, radio, and x-ray light hit the Earth. The visible front was soon detected by astronomers in Chile, and astronomers worldwide mobilized to capture the rare event.

The neutrinos were the product of a supernova explosion, the first in almost four centuries bright enough to be seen by the naked eye from Earth. It happened, actually, some 160,000 years ago in the Large Magellanic Cloud, a small companion galaxy to our own. The last bright supernova in our own galaxy erupted before Galileo turned his telescopes to the sky. In 1987, telescopes on the ground, and in space, greeted the event.

Supernovae are spectacular events and give us important clues to answers to some very fundamental questions, such as, "Where do we come from?" More precisely, where do the materials we are made of come from? We know that the big bang produced only the lightest elements in the universe: hydrogen, helium, and a bit of other light elements. Most of the rest—iron, oxygen, carbon, etc.— were synthesized in the nuclear fusion reactions in stars. But even in those terribly hot places, not all elements could be synthesized. The remainder, the heaviest of the elements from iron and beyond, were forged in the catastrophic collapse and explosion of massive stars—in supernovae—and neutrinos are the products of these violent reactions. A few dribble out of the sun under normal conditions, as hydrogen fuses into helium at the sun's center. But a supernova produces enormous quantities

Stars spend the vast majority of their lives behaving very nicely, quietly converting hydrogen into helium through thermonuclear fusion processes, and this creates energy that balances the inward press of gravity. But when the hydrogen available for fusion is exhausted, the star become unstable and goes through a series of steps to adjust its structure until other sources of fusion energy are triggered. Each one of these steps builds up the heavier elements up to iron and releases great amounts of energy, which keeps the star from collapsing. But when the stellar core becomes enriched with iron—which happens very rapidly as each succeeding stage of fusion cycles faster and faster—something really nasty happens. The iron requires more energy to fuse than the fusion process releases, and the star is left hanging as a vast, bloated sphere of gas, driven inward by its own gravity.

When fusion in the iron-enriched core shuts off, the outerlying mass of the star collapses the core. Electrons smash into protons to produce a violently dense sea of neutrons, and a swarm of neutrinos are released. The collapsing outer shell cannot penetrate this neutron wall, and so rebounds with an enormous acoustic shock wave back into the star. These shocks reheat the outerlying gases sufficiently to produce explosive processes that quickly form heavy elements

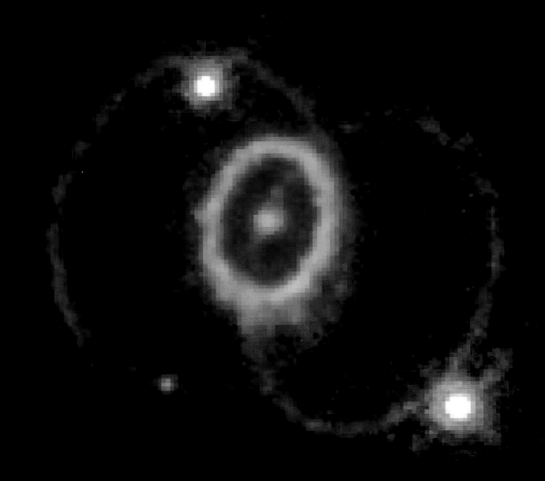

in the expanding blast wave. The neutrinos stream out into space at the speed of light, unhindered by intervening matter.

The 19 neutrinos that were caught in 1987 are clues to the star's core collapse. The elusive neutrino can be thought of as the messenger agents in the evolutionary process driving the universe. Supernovae can tell us about our elemental origins.

Neutrinos are the closest something can be to being nothing and still be something.

They have been seen in particle accelerator experiments and from the sun, but Supernova 1987A provided our first glimpse of neutrinos from outside the solar system. It proved that neutrinos could last for 160,000 years—their travel time from the Large Magellanic Cloud—and provided scientists with more clues about the origin of the universe's elements and the processes of transformation that have built our universe.

—David DeVorkin

Glowing rings of gas puff outward from a supernova that flared into view in 1987. The Hubble Space Telescope took this visible-light image seven years later. The supernova occurred in the Large Magellanic Cloud, a nearby dwarf galaxy.

LEARNING FROM THE PAST: HISTORICAL SUPERNOVAE

to provide strong evidence that a supernova had been sighted in 1664. In my source, a late Korean historical compendium titled *Chungbo Munhon Pigo,* a bright star was reported as having been seen for two months in the Scorpio region that year. This seemed an exciting discovery! However, when I consulted the original, very extensive chronicle (in the *Yijo Sillok*), I could find no parallel record. I eventually realized that the account in the *Chungbo Munhon Pigo* actually related to the supernova of 1604. The error had arisen because of the frequent use of a 60-year cycle in East Asian history. A careless scribe had written 1664 instead of 1604.

Once I had acquainted myself with the most useful—and reliable—sources of information about supernovae from centuries ago, my next major task was to build up a collection of copies of early East Asian star charts in order to locate the positions of the "new stars" as accurately as possible. Chinese, and later Japanese and Korean astronomers, usually recorded the positions of guest stars relative to specific star groups. However, the constellation figures envisaged in the Orient bear almost no relation to those of the Western world. In ancient China the night sky was divided into as many as 283 star groups, containing a total of about 1,460 stars. For comparison, the star catalog of Claudius Ptolemy, the great

Chinese celestial charts prove invaluable in pinning down the locations of historical supernovae, but deciphering them is a challenge. They can depict nearly 300 constellations, few of which resemble those of other cultures, and the stellar brightnesses shown on the charts do not always match those seen in the sky.

Greek astronomer of the second century, listed only 48 constellations containing a total of 1,022 stars. There are very few instances where the two groupings are similar apart from the Big Dipper (Beidou, the Northern Dipper in East Asian astronomy), Orion (Shen, the Triad), and the Tail of the Scorpion (Wei, the Tail of the Dragon).

Throughout East Asia, guest stars were recorded mainly because of their astrological significance. The Chinese empire was regarded as a microcosm of the universe; each star group had its equivalent within the empire and was believed to exert an influence on it. As a result, we find stars or star groups with names such as "Great Emperor," "Empress," "Secretaries," "Celestial Fields," "Celestial Kitchen," "Celestial Prison," and even "Wailers" and

This unusual star chart shows both the traditional 28 Chinese lodges and the symbols of the Western zodiac. Taurus is missing from the defaced area to the right. This Liao dynasty tomb ceiling chart dates from 1116.

LEARNING FROM THE PAST: HISTORICAL SUPERNOVAE

"Weepers" adjacent to "Burial Place." A guest star appearing in or near any of the 283 recognized star groups was often regarded as presaging some event involving the corresponding personage or object in China itself.

A guest star was seen in China for about five months in A.D. 369. The official history of the time, the *Hou Hanshu* (*History of the Later Han Dynasty*), records that the star appeared at the "Western wall of Ziwei." The text adds: "The Interpretations say, 'When a guest star guards Ziwei, it means assassination of the Emperor by his subjects.'"

Japanese and Korean astronomers/astrologers closely followed traditional Chinese mapping of the stars, but predictions based on their observations were applied to matters in their own countries. Individual horoscope astrology, as practiced extensively in the Western world, never made any significant headway in East Asia, but there can be no doubt that without the impetus provided by celestial divination, many astronomical events would neither have been observed nor recorded down the centuries.

Oriental records specify the positions of guest stars in two basic ways: the star group in or near which the object appeared, and the appropriate lunar lodge, or *xiu*, in which it appeared. From ancient times the circle of the sky was divided into a series of 28 of these lodges, which were unequal zones radiating outward from the north celestial pole. Eastern astronomers sometimes measured the right ascension (similar to longitude on the Earth's surface) of a new star eastward from the boundary of the appropriate lunar lodge. The following report of the supernova of 1181 in the *Songshi,* the official history of the Song dynasty in China, contains examples of both of these ways the Chinese used to document a star's position.

> *Chunxi reign period (of emperor Xiao Zong), eighth year, sixth month, day jisi [Aug 6, 1181].*
> *A guest star appeared in Kui lunar lodge, invading the stars of Chuanshe until the following year, first month, day guiyou [Feb 6, 1182], a total of 185 days. Only then was it extinguished.*

Kui, meaning "Stride," was the 15th of the 28 lunar lodges and was some 15 degrees in width; Chuanshe, or "Guest Houses," was a star group in the Western constellation of Cassiopeia, within the coordinate range covered by the lunar lodge of Kui. By incorporating other celestial locations for the guest star, as described independently by court astronomers from North China (then a separate empire) and Japan, it is possible to deduce its position to within a few square degrees. Modern catalogs of supernova remnants observed at radio and x-ray wavelengths shows that, in this part of the sky, supernova remnants are thinly scattered. In fact, only one of these objects could possibly correspond with the new star seen in 1181—the remnant known as 3C 58, the one I found during my doctoral research. This remnant is an active source of electromagnetic radiation, and 800 years is within the estimated range of its age.

The date of the above observation is expressed on the traditional Chinese lunar calendar, which was later adopted in both Korea and Japan and has remained largely unchanged since the second century B.C. However, in each of the three countries, years were usually counted from the start of the appropriate ruler's reign. Most years contained 12 lunar months, or about 354 days, but every two or three years a 13th month was added to keep the start of the year in step with the seasons. At some point in very ancient history, a 60-day cycle that ran independently of the lunar calendar was introduced. Sometimes a 60-year cycle was also used, as it is today; for instance, the year 2001 is the 18th year (*xinsi*) of the present cycle.

The new star of 1006 is probably the most brilliant supernova on record. Many descriptions of the star are preserved from both China and Japan, and also some from the Arab lands and Europe. However, the most detailed observations are from China. Chinese astronomers compared the brightness of the new star, which was first seen on a date equivalent to May 1 in A.D. 1006, with that of the half-moon. They described the star as "huge," saying, "its bright rays were like a golden disk," and that "one could really see things clearly by its light." Arab writers also compared the star with the moon, remarking that "the sky was shining because of its light," while others at the monastery of St. Gallen in Switzerland described it as "dazzling the eyes."

The supernova of 1006 was probably the brightest ever recorded. One observer described it as "dazzling the eyes." This x-ray image reveals what remains of it a thousand years later.

The position of the supernova was measured by Chinese astronomers as 3 degrees in Di ("Root"), the third lunar lodge (of width 15 degrees). Two other star groups mentioned in the Chinese records were Kulou ("Depot Tower") and Qiguan ("Cavalry Officer"). A study of medieval Chinese star charts shows that Kulou and Qiguan were adjacent to one another in the Occidental constellations of Centaurus and Lupus. The new star was thus in the far southern skies. Japanese records confirm a location near Qiguan. The Chinese and Japanese official observatories were both close to latitude 35 degrees north, so the star was well placed for visibility from these locations. However, much farther north, at St. Gallen, the star only grazed the mountainous southern horizon. The mere fact that the star could be seen from there sets a useful limit on its declination (a celestial coordinate similar to terrestrial latitude). A further measurement reported by Ali ibn Ridwan, a Cairo physician, gives the star's position on the zodiac as the 15th degree of Scorpio.

With this information, modern astronomers have been able to locate the remnants of the supernova, called PKS 1459-41. Radio and x-ray observations of PKS 1459-41 indicate that an age of about 1,000 years is realistic. The only other known supernova remnant within about 10 degrees of the recorded position is the very ancient Lupus Loop, which has an estimated age of about 10,000 years.

Asian and Western observers followed the star throughout the summer of A.D. 1006, but it was lost to view in the evening twilight during September of that year, and only Chinese astronomers ever reported it again. At Bian, the capital of China (modern Kaifeng), the star

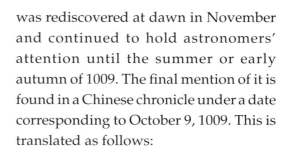

was rediscovered at dawn in November and continued to hold astronomers' attention until the summer or early autumn of 1009. The final mention of it is found in a Chinese chronicle under a date corresponding to October 9, 1009. This is translated as follows:

> *Emperor Zhenzong ordered that from this day the Suburban Sacrifice should be arranged for the Zhoubo star located at Di lunar lodge...for ever. This met with the approval of the Hanlin academicians and the astronomers.*

Clearly the supernova of 1006 left a profound impression on many eyewitnesses. It seems ironic that modern astronomers, with all their scientific expertise and array of instruments, have yet to witness anything like such a brilliant stellar outburst in the Milky Way, but such events occur randomly in time and it may be many years before a similar opportunity arises. In the meantime, astronomers have little option but to make the best use of the historical observations of galactic supernovae.

171

During its violent death throes, a supernova can radiate the energy of a billion suns and rival the brightness of an entire galaxy, before it slowly fades from view. This one at left appeared in 1994 in the nearby Virgo galaxy cluster. The brightness of this particular type of supernova can be used to gauge galaxy distances.

THE LARGE-SCALE STRUCTURE OF THE UNIVERSE

Margaret J. Geller

On family trips, my sister and I sat in the back of our parents' roomy late-1940s Pontiac. There was plenty of space to spread out a road map. My eyes repeatedly darted from the map to the scenery and back. Again and again I identified the scenery and the road markers with the markings on the pastel map, with its occasional grimy marks from my eager young fingers. My romance with maps began on these childhood trips.

Unlike the exquisite old maps reproduced so widely in books and calendars, modern maps contain little hint of the mystery and magic of small roads and towns. Nonetheless, for me they are still irresistible invitations to discover.

Drawers at home are full of maps that tell part of the story of my life and my exploration of a small portion of the world.

Approaching the Milky Way from a distance of 100 million light years, a computer-animated space voyage reveals the large-scale structure of the universe: vast concentrations of galaxies and empty, dark voids.

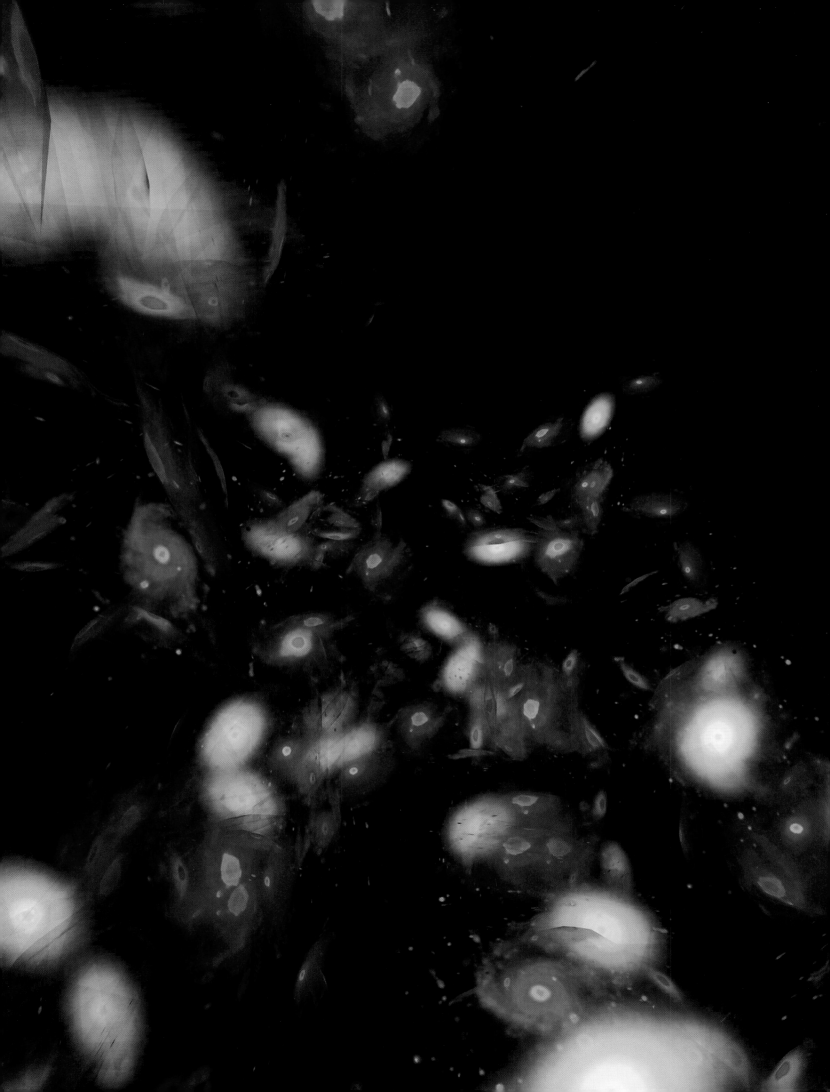

My office is also full of maps, but they are of a very different kind: I had a hand in making the maps in my office. They display slices of the nearby universe covering distances of hundreds of millions of light years.

I became a scientist because my father was a scientist. He showed me the profound connection between maps and science, between maps and the geometry of nature. With him as my guide, I peered through a microscope into the wonder-world of hexagonal snowflake designs. Through the lenses of geometry and physics, there is a universe in a snowflake. In childhood conversations I absorbed the idea that there is a connection between the patterns we see in nature and the laws of physics.

I have always liked big questions. Like many scientists before me, I map the universe (in truth, small portions of the universe) to discover what it looks like and how it came to be as we observe it today. Fortunately, my maturation as a scientist coincided with a time when advances in technology made it possible to begin to answer these questions.

Not so long ago, when my parents were young children, there was a revolution in scientists' view of the universe. In 1929, the year of the stock market crash, Edwin Hubble demonstrated the expansion of the universe. He showed that galaxies appear to recede from us with velocities proportional to their distance. This relation was just the one predicted in evolving models of the universe derived from Albert Einstein's new theory of general relativity. In that theory space itself is dynamic; the stretching of the space between the galaxies, the expansion of the universe, explains why galaxies appear to fly away from us (and from every other observer in the universe).

174

Today we use Hubble's relation to map the way galaxies, including our own Milky Way, mark the distribution of matter in space. Pictures of the sky are the first step toward a map. For decades, photographic plates were the recording medium for these images. During the 1950s, National Geographic supported the use of the 48-inch Schmidt telescope on Mount Palomar in southern California for photographing the entire northern sky. During the 1960s, Fritz Zwicky and his collaborators examined the roughly 1,000 sky survey plates and identified more than 30,000 galaxies. We now know that these galaxies are mostly within a billion light years of the sun, our neighborhood in the visible universe, which extends for some 15 billion light years around us.

In the 1970s, when I was doing my graduate work at Princeton and my postdoctoral work at the Center for Astrophysics, the Zwicky catalog became a cornerstone for maps of the nearby universe. These two-dimensional maps are revealing and frustrating at the same time. They enabled a statistical description of clusters of galaxies that extend for millions of light years and contain hundreds to thousands of objects. But the catalog provides only the longitudes and latitudes of the galaxies. Galaxies appearing near each other on the sky can be at very different distances. Long before the technology to measure it came of age, the need for the third dimension—the distance to the galaxies—was obvious.

Astronomical distances are notoriously difficult to measure, but there is an easier approach to placing galaxies on a grand three-dimensional map. Conveniently, Hubble's law links the expansion velocity with the distance to a galaxy; it says that distance is simply proportional to velocity. The astronomer's route to measuring the distance is to measure the velocities, or, more precisely, redshifts. A spectrograph attached to the telescope spreads the light from a galaxy out into its colors. The resulting spectrum shows bright and dark lines, the signatures of elements such as hydrogen, oxygen, nitrogen, sodium, and calcium. In laboratories on Earth we can easily observe the spectra characteristic of these elements and measure the precise colors of the lines. In the spectra of galaxies we see the same patterns of lines, but they are stretched to longer wavelength. We call this stretching a redshift. The more distant the galaxy, the greater the redshift; the redshift is proportional to the velocity and thus to the distance. By measuring redshifts of many galaxies, we escape the two-dimensional world of pictures of the sky to construct a three-dimensional map of the universe.

In Hubble's day, measuring redshifts was a slow, tedious process. Modern technologies have made these measurements rapid and routine. We have larger telescopes and better optics to collect, reflect, and transmit the ancient photons (particles of light) that reveal the secrets of the structure of the universe. We detect these precious photons with sensitive CCD (charge-coupled device) detectors (similar to the devices in digital cameras) rather than photographic plates. The digital data travel over the Internet from telescope to laboratory. Fast computers extract the redshift from the raw spectrum. Many people—scientists, engineers, programmers, and technicians—design and build the instruments, develop computer codes, and make and interpret the observations. All these talented, curious people contribute to making ours the age of mapping the universe.

In the early 1980s, two observations aroused my curiosity about the largest patterns, the continents and oceans, of the universe. Maps of the sky much deeper than the ones made by Zwicky showed thin filamentary concentrations of galaxies stretching hundreds of millions of light years across the sky. Valerie de Lapparent, Michael Kurtz, and I examined these apparent structures and argued that they were artifacts of the methods used to construct the galaxy catalogs. In spite of this disappointing result, we were intrigued by the possibility that galaxies marked out large patterns in the universe. We were not at all alone. Several groups of astronomers were measuring redshifts of galaxies; their maps revealed dark regions nearly devoid of galaxies. The most intriguing of the dark regions uncovered in these early explorations was the "void in Boötes." It was fascinating because it was enormous, a dark region nearly 300 million light years across.

Large patterns like the filaments in the deep map or the void in Boötes weren't supposed to exist. The picture of the universe most astronomers (including me) carried in our minds was simpler and, I now think, more boring. Clusters of galaxies are obvious in two-dimensional maps of positions of galaxies on the sky. These systems, held together by gravity, are a few million light years across and contain hundreds to thousands of galaxies comparable in mass with our own Milky Way. A few clusters may form a supercluster extending for a few tens of millions of light years. Many thought that clusters and perhaps superclusters were the largest structures in the universe. An artist's conception of the universe might have shown clusters and occasional superclusters distributed within a random sea of additional galaxies. There was no room in this picture for a dark region like the void in Boötes. Like many, I thought there must be some mistake—not in the standard picture but in the new observations.

Looking back, I am surprised at my own conservatism. The more I have done scientific research, the more I have realized what a conservative process it is. It is remarkably hard to depart from cherished truths, even if there is precious little evidence supporting them. By 1984 I had recognized that no one really knew the arrangement of galaxies. I suggested to John Huchra that we make a three-dimensional map of a slice of the universe. This project became part of the Ph.D. thesis research of my student, Valerie de Lapparent.

To map our slice of the universe we used the 1.5-meter telescope operated by the Smithsonian Institution on Mt. Hopkins, Arizona, to measure a redshift for each galaxy in a strip stretching from east to west across the northern sky. An analogy with mapping the Earth makes the rationale for the strip clear. Suppose you had never seen the Earth but wanted to know whether it had any large features—continents and oceans. To answer the question you are restricted by instruments and time to a small contiguous portion of the Earth, let's say one one-hundred-thousandth of its surface (about the area covered by the state of Rhode Island). If you survey a patch, you will most likely land in the ocean and fail to answer the question. Surveying a thin great circle around the Earth in almost any orientation reveals that the Earth has two kinds of structures, continents and oceans, both big. There are a few great circles that pass though ocean only, but you would have to be unlucky to choose one of those. Of course, when we choose a slice of the universe, we have to hope that the one we choose is typical. Fortunately, we can test that assumption by mapping others.

Even in 1985, the 1.5-meter telescope ranked as a small instrument, but it was big enough to explore the nearest 700 million light years of the universe and to uncover other patterns like the void in Boötes . . . *if* they were there. Night after night for a year, we pointed the telescope at one galaxy after another to acquire the spectrum and measure a redshift. Because we did not expect a remarkable result (no one ever does), John and I did not encourage Valerie to plot the data before they were all in hand.

Seeing the data for the first time gave me goose bumps. There before us was an enormous pattern clearly delineated by the thousand galaxies in the survey. We were, I suppose, fortunate that nature was not subtle. In hindsight other surveys before us contained evidence of these patterns, but the hints were too subtle to overturn the generally held belief in their absence. Our slice of the universe shows voids just like the one in Boötes, surrounded or nearly surrounded by galaxies. We suggested that the galaxies mark a bubble-like structure: The dark voids are the interiors of the bubbles; the galaxies mark the thin surfaces around them.

By 1989, John Huchra and I had used the 1.5-meter telescope to more than quadruple the number of redshifts in our map. As the first slice suggested, this more extensive map showed that many galaxies occupy thin walls. The Great Wall, the prototype for these enormous structures, extends across our survey for more than 800 million light years. At the time of its discovery, the Great Wall was the largest structure anyone had ever seen. Now we know that great walls are a common feature of the universe.

During the 1990s two technological developments again accelerated the pace of redshift measurement. High-tech companies developed methods for producing larger, more sensitive CCD detectors. Driven by the explosion in the communications industry, fiber optics

came of age. Scientists used these technological advances to enable simultaneous observation of first tens and now hundreds of galaxies at once.

Larger telescopes peer much deeper into the universe than our humble 1.5-meter. The deeper we look, the farther back we see in time. The deeper we look, the more galaxies we see in a square degree of the sky (for calibration, a square degree is about five times the area covered by the moon). Our slices of the universe reach to a distance of about 700 million light years; to this depth there are only one or two galaxies per square degree of the sky. A six-meter or larger telescope, for example, can readily return redshifts for galaxies at distances of five billion light years or more; at this depth there are thousands of galaxies per square degree. The typical galaxy is only two-thirds the age of the Milky Way.

The first step in mapping the universe is still locating the galaxies in images of the sky. Although some of these redshift surveys are still based on catalogs derived from astronomical photographic plates, digital surveys of the entire sky are underway. For the largest mapping projects (redshift surveys) in progress now, the coordinates for galaxies become the positions of optical fibers in the focal plane of the telescope. Just as in a camera, the focal plane of a telescope is where the focused image of the sky appears. Each of several hundred optical fibers, one for each galaxy observed, carries the ancient light from a galaxy to the

In this map of a slice of the universe, each of the 5,000 dots represents a galaxy. The map reaches from the Milky Way at the lower vertex to a distance of two billion light years at the upper curved boundary. The galaxies trace a bubble-like pattern.

THE LARGE-SCALE STRUCTURE OF THE UNIVERSE

HOW BIG
IS BIG?

A favorite game for teachers, students, and families alike when thinking about astronomical distances is to map out the size of the solar system on a playground, in a mall, or just about anywhere convenient. Some towns and cities have created scale models to compare their familiar spaces to distances between the Earth and sun, moon or planets. And recently, a new exhibition opened on the National Mall that repeats the exercise in grand scale. The sun, about the size of a large grapefruit, sits at the northeast corner of the National Air and Space Museum, and Pluto, a mere grain of sand, is some 0.6 kilometers away at the northeast corner of the Smithsonian "Castle"—the distinctive dark red administration building. The Earth, slightly larger than a black mustard seed, is about 16 meters from the sun, and most of the solar system is spread out along Jefferson Drive in front of the Air and Space Museum. But we need to go farther, much farther to include the distances to galaxies.

At this scale, about one ten-billionth of the size of the actual solar system, a light year would be about 1,000 kilometers long, and the nearest star system, Alpha Centauri, would be 4,300 kilometers from the National Mall, just a bit greater than the distance from Washington to San Francisco, California, as the crow flies. The center of the Milky Way galaxy (32,000 light years) would be almost as far as the circumference of the Earth straightened out into a line. And the nearest major galaxy, M31 in Andromeda (2.2 million light years away) would be two-thirds the distance to the Moon! At this point it makes no sense to go any further because our yardstick is as incomprehensible to us as the actual distances involved.

On the scale of mankind, or even the sun or the solar system, the universe is a big place. But all this is apparent. There is just as much "space" in the microcosmic world. The limits of large and small are limits of our perception only. To comprehend these limits, we have commonly taken a "Powers of 10" approach. The first known attempts at this appeared about 100 years ago in literature. Various American museums, such as the American Museum of Natural History in New York, are known to have tried out three-dimensional models in the 1920s and 30s. Since then, many books, models, films, videos, and now computer interactives have given it a try.

All of these efforts require that the viewer imagines vastly different scales of existence in three dimensions. If a cube one-meter on a side in the first scene becomes a cube 10 meters on a side in the next scene, the new volume is 1,000 times larger than the first volume. This is comprehensible in a three-dimensional exhibit, but what about in a book like you are holding right now? This expansion by 1,000 is accomplished by moving only ten times the distance from the origin than you were before. Unless you can think in three dimensions, what you perceive, however, is not an expansion in volume, but only a linear expansion in distance. Still, we try to place these perspective drawings in books, and cross our fingers that our readers will ponder their significance in our three-dimensional world.

—David DeVorkin

As our cosmic perspective has evolved, so has our understanding of the extent of the universe. As larger telescopes enable us to study objects at greater distances from the Earth, our picture of the universe has changed from a starry sphere enclosing the Earth to a vast web of galaxies extending for billions of light years.

THE LARGE-SCALE STRUCTURE OF THE UNIVERSE

spectrograph on the floor of the observatory. The spectrograph spreads the light out into its colors. Large CCD detectors record the resulting several hundred spectra. When we mapped our first slice, we measured about 25 redshifts during a good night. Today a telescope with a modern fiber instrument readily returns 2,500 redshifts per night for galaxies at ten times the distance (or more) to the ones in our survey!

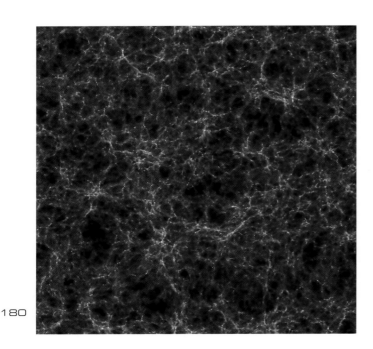

A computer simulation shows that a slice of the youthful, billion-year-old universe has little structure. The nascent voids and filaments are barely visible.

Teams of scientists based in Australia and in the United States have undertaken ambitious mapping projects that rely on the new technology. The Australian project goes by the name of the instrument that enables it, the 2DF, for 2-degree field. The 2DF on the 4-meter Anglo-Australian telescope returns nearly 400 redshifts at a time for galaxies distributed across a 2-square degree region of the sky. The 2DF maps now include 175,000 galaxies. The patterns in these impressive maps are similar to the ones we discovered. Because the maps are more extensive than ours, they contain many dark voids, along with a multitude of thin walls and filamentary structures where the galaxies are.

The 2DF map is more than three times as deep as ours. A consortium of universities in the United States has undertaken an even more audacious project; they plan to image a quarter of the sky digitally and to acquire spectra for one million galaxies.

The first slices of these large surveys give the same message as the ones before them: Dark voids, thin walls, and filaments define the bubble- or sponge-like tapestry of our neighborhood in the universe.

The wealth of new data enables us to write our detailed address in the universe. Taken together they show that we orbit an ordinary star on the outskirts of an ordinary galaxy in an ordinary group of galaxies on the outskirts of an ordinary cluster of galaxies in an ordinary patch of the ordinary patterns in the universe. What perhaps departs from the ordinary is our ability to ask increasingly detailed, precise questions about the nature of our grandest surroundings and their history. Among these questions, the most profound and frustrating is the quest to determine the nature and distribution of the matter in the universe. The spectacular maps show us the distribution of the material which emits light. The relative motions of galaxies we infer from these maps tell us that most of the matter in the universe, more than 90 percent of it, is dark. This conundrum has been with us, unresolved, for nearly 70 years.

Knowing the nature of the dark matter is crucial for a complete understanding of the

180

formation of galaxies. The cosmic microwave background radiation (discussed by David Wilkinson elsewhere in this volume) that pervades the universe carries our earliest glimpse of the clumping of matter in the universe. This radiation has a remarkably uniform temperature of 2.7 degrees Kelvin. Tiny fluctuations in this radiation of one part in 100,000 are a kind of picture of the lumps and bumps in the distribution of matter when the universe was a mere 300,000 years old. If the surface of the Earth were as smooth as the early universe, the highest point on Earth would be 60 meters . . . boring!

In remarkable agreement with the analysis of redshift surveys, study of the miniscule fluctuations in the cosmic background radiation tells us that only about 10 percent of the matter in the universe is the normal baryonic stuff that makes up the objects we observe: stars, planets, and human beings. The other 90 percent is something still mysterious and dark.

Between the microwave background fluctuation epoch and today, stars and galaxies formed. Life to observe and wonder about them evolved. Fortunately, throughout those nearly 15 billion years, the universe is transparent. Ancient photons travel to us essentially unimpeded, carrying the story of the history of the universe. We collect and detect this ancient light and read the tale.

As the universe evolves, patterns develop. The simulated 15-billion-year-old universe looks like the observed slice with clearly defined voids, sheets, and filaments.

To interpret the data, we assume that the laws of physics we discover in our laboratories on Earth apply at every place and every time in the universe. As the Nobel laureate Sheldon Glashow recently commented, these physical laws apply "everywhere and everywhen." I have always regarded the remarkable success of this approach as part of the wonder and romance of science. It is easy to imagine a species able to appreciate the beauty of the natural world but unable to investigate it in detail or to develop mathematical models that have predictive power.

Sophisticated computer models based on our knowledge of fundamental physics are important tools for predicting what we might observe when we explore the distant young universe. Like many others, I have used these simulations for planning large observational projects. We pretend that we are observing the simulations and extract "data" that tell us what to expect (if the model is correct) when we observe the real universe. When the observations are complete, the simulations provide a standard for interpretation and understanding. If the data and models match, we understand the underlying physical process. If

they don't match, we have to figure out what is missing from the models or the data.

In the largest simulations, astrophysicists follow the evolution of a distribution of a billion particles interacting gravitationally in an expanding universe. The tiny lumps and bumps revealed by the radiation background are the seeds for the galaxies, clusters of galaxies, and larger patterns we observe. The background fluctuations specify the starting point for the models. In an expanding universe, gravity grows both lumps and holes. Simulated universes

Large telescopes reveal the distant universe. Most of the fuzzy colored objects in this image of the sky are galaxies. Because of the redshift, distant galaxies appear redder than nearby ones. The most distant red objects are more than ten billion light years away and they are only a few billion years old.

bear a striking resemblance to the real cosmos.

These very successful simulations still contain many untested assumptions. The nature of the dark matter is perhaps the most profound of these. The unresolved issues that fascinate me most center on the behavior of the baryonic matter. Perhaps this focus is an anthropocentric view: We are made of baryonic matter and the exquisite objects we observe directly are made of it too. I am partial to the universe we *can* see. I would like to know how stars form in galaxies. I wonder how galaxies manage to become spirals like our own Milky Way or giant ellipticals like the ones that inhabit the center of the Coma cluster. I want to see how the patterns revealed in our first slice formed. My dream is to see the real universe movie that shows the development of our universe, not a simulated one, from an age of 300,000 years to the present.

The universe is a time machine. When we look out in space, we look back in time. When we observe a galaxy a billion light years away, we see the galaxy as it was a billion years ago. The farther we look, the younger the objects we observe. With the right instruments, nearly the entire history of the universe is there for us to see. A famous image taken with the Hubble Space Telescope, the Hubble Deep Field, shows us young galaxies only a few billion years old. Redshifts for many galaxies in this tiny patch of the young universe show that even then, there were walls rather similar to the ones in our first slice. The image is a measure of our reach and an icon of our time. In 2009, with the launch of the Next Generation Space Telescope, we will have images of even younger objects.

In spite of this prospect, I am unlikely to realize my dream fully. On one snowy day a couple of years ago, I calculated how long it would be before we have a redshift measurement for all the galaxies in the visible universe. If technology continues to advance at the rate it has since the 1970s, we will have a map of the entire visible universe by the year 2100.

Fortunately for me, a partial map is enough to answer plenty of interesting questions. Next year, my colleagues and I plan to begin mapping a slice of the universe as it was at about a two-thirds of its age, or about five billion years ago. We will use a robotic fiber instrument, the Hectospec, designed by Dan Fabricant, mounted on the newly upgraded MMT with a 6.5-meter monolithic mirror at its heart. In the Hectospec, two robots, Fred and Ginger, place 300 fibers at the accurately measured positions of faint, distant galaxies. They do their work in less than five minutes, moving at speeds up to a meter per second. In about an hour we will obtain the redshifts of nearly 300 galaxies at once. These middle-aged galaxies are typically more than five times farther away than the galaxies in our first slice.

The Hectospec is one of several sophisticated instruments designed to map the history of the universe. An instrument called Deimos, designed for the ten-meter Keck telescope, will map the adolescent universe. Taken together these instruments will return data that answer the question, How did the universe come to be as we see it today? All of these instruments employ some of the finest, most sophisticated technology of our age to explore the largest physical system we can know. In the Hectospec, fibers only twice the diameter of a human hair carry light from galaxies a hundred thousand light years across and five billion light years away.

I, for one, can't wait to see the results of these surveys. We expect to see a less clustered distribution of galaxies. We know that we will see more star-forming galaxies in clusters than we see in the nearby universe, but by comparing the two epochs we hope to understand how galaxies in clusters evolve and how the clusters themselves grow. Computer models give us the clues, but there is no substitute for the measurement. From an aesthetic point of view, for me, at least, there is exquisite beauty in the natural world that simulations cannot match. On the other hand, the use of physical laws to understand makes the beauty richer. For me there is a deep satisfaction and a floating feeling of peace in knowing that I can play a part in this understanding, and that I have been the first to uncover an enduring beauty of nature.

I would like to see a picture of the entire visible universe just as I would like to visit all the places on the maps in my drawers at home. Unsatisfied curiosity is vaguely unsettling, especially to a scientist who likes the big picture. I suspect that if the enterprise of scientific research remains as we know it today, the grand-students of my students will see such a picture. Nonetheless, some small portions of the universe are as familiar to me as my favorite places for summer vacation. I feel privileged to have played a part in one of the grand and challenging explorations of our time.

Complex instruments on large telescopes enable 3-D mapping of the distant, younger universe to compare with models. In the Hectospec, an instrument for measuring galaxy redshifts, fast robots position each of the 300 optical fibers shown here at the location of a distant galaxy.

THE LARGE-SCALE STRUCTURE OF THE UNIVERSE

WHY DOES AN OPTICAL ASTRONOMER STUDY SOMETHING SHE CANNOT SEE?

Vera Rubin

We live in a universe that is breathtakingly beautiful, incredibly large, and enormously complex. Even more remarkable, we have evolved with brains that can study and understand some secrets of this mysterious place. Our progress in understanding our universe is accelerating, yet there is so much more to be learned.

Back in 1960, we knew the general structure of our home galaxy, the Milky Way. It is spiral and disk-like, some 100 billion stars bound together by gravity and orbiting a distant center. We also knew that the stars in this disk are not distributed at random, but flattened to a "galactic" plane, in which our solar system is located. When we look through the plane we see the bright band of millions of stars that gave the Milky Way its

Only available starlight was used to photograph the brilliant band of the Milky Way, our home galaxy, seen here illuminating the four-meter Blanco telescope at the Cerro Tololo Inter-American Observatory in Chile. Our two close galactic neighbors, the Large and Small Magellanic Clouds, appear at the far left.

WHY DOES AN OPTICAL ASTRONOMER STUDY SOMETHING SHE CANNOT SEE?

name. This is one of the greatest views on Earth, now tragically lost to most Americans by the glow of city lights.

Most ancient civilizations told stories to explain this bright path across the dark sky. It was not until 1609—when Galileo placed a small lens at one end of a tube and his large brain at the other—that we learned the Milky Way is composed of individual stars. Galileo's genius lay not only in discovering things never before seen in the sky, but also in correctly interpreting what he saw.

But what about what we cannot see? Our galaxy contains more than stars. That galactic plane also churns with gas and dust so dense that our view of more distant objects is obscured. From our position in the galaxy we cannot see all the way to the center with visible light; we must use instruments that allow us to see the infrared part of the spectrum—and we must place those instruments above Earth's atmosphere. Turning our backs to the galactic center, we can use a telescope to look away from the plane of the Milky Way, and then we see external galaxies, each an independent agglomeration of billions of stars.

Gravity makes a galaxy. Like the sun's ability to hold the planets in our own solar system in their orbits, the gravitational center of a galaxy captures billions of orbiting stars. That same gravity also retains the gas shed by dying stars and recycles it into the next generation of stars.

Because telescopic views of other galaxies seen nearly edge-on resemble the view we have of the Milky Way, we are convinced we understand the most prominent features of our galaxy. When we view a distant galaxy face-on rather than from its edge, we see a bright central region formed from the combined light of billions of stars. The spiral structure of these galaxies gets fainter with increasing distance from the galaxy's nucleus. Some galaxies show a tight, regular spiral pattern. Some show a more ragged spiral pattern with open "arms." Some have a large central bulge, and others have almost no such bulge.

As a young assistant professor at Georgetown University in Washington, D.C., in the early 1960s, I was fascinated by the variety in galactic shapes—and by the outer, mostly unknown regions of the Milky Way. Radio astronomers were beginning to investigate the orbital motions of the cold, dark, and electrically neutral gas that fills the enormous spaces between stars. I hoped that, by measuring the speed and direction of stars orbiting the visible galaxy, we might find clues as to why galaxies came in so many forms. Later, I also came to realize that I had stumbled upon an area of study that was relatively untouched, where I could work at my own pace, without competition from other astronomers. Like stars themselves, my fellow astronomers seemed clustered at the center of the galaxy, rather than its fringes.

As a class project, my students and I derived the orbital motions of about 1,000 stars in the Milky Way, located at distances farther from the center of our galaxy than the sun. From star catalogs we gathered data concerning the velocities of young, bright stars that had formed some millions of years ago—mere days by astronomical standards—from interstellar gas in the galactic plane. The findings were surprising: The velocities of the stars did not decrease as their distance from the galactic center grew. If the greatest mass of the galaxy was at its center—and it certainly looks that way through an optical telescope—then, according

to laws determined 400 years ago by Johannes Kepler, stars at a greater distance from the center of mass should travel slower.

As a mathematician working in Tycho Brahe's magnificent observatory, where the finest tables of planetary positions were being created, it was Kepler's job to refine "planetary theory," or the mathematical models that predict how the planets, sun, and moon move among the stars. Kepler knew, for instance, that Mercury, the planet closest to the sun, orbits very rapidly, and that Saturn, a planet distant from the sun, orbits very slowly. What Kepler found from Tycho's highly accurate data was that the planetary orbits, particularly that of Mars, were not circular, as had been assumed since antiquity, but elliptical. In formulating his laws of planetary motion, Kepler laid portions of the foundation for Isaac Newton's theory of universal gravitation.

Newton taught us that the sun's gravitational force falls off with greater distance from from it. The planets are actually falling toward the sun, but their forward motion is so great

From afar, the Milky Way looks much like this edge-on view of spiral galaxy NGC 891. The bulging core and surrounding disk are so distant that we cannot discern individual stars. The single stars visible in this image are in the Milky Way.

WHY DOES AN OPTICAL ASTRONOMER STUDY SOMETHING SHE CANNOT SEE?

that instead of reaching the sun, they fall "around" the sun. In other words, they orbit it. Planets nearer to the sun—the center of the solar system's mass—orbit faster than those farther away. Using their distances from the sun and their orbital periods, we can actually calculate the sun's mass.

If we applied similar reasoning to galaxies, we would presume that the mass within a galaxy could be determined from the orbital velocities of stars or gas at successive distances from the center of that galaxy. Most of what we see in a spiral galaxy—its overall brightness, or luminosity—comes from its central portion. It becomes dimmer with increasing distance from the center, so we might think that this luminosity is a reliable predictor of how much mass a galaxy contains and where most of that mass is located. So, using this predicted distribution of mass and Newton's laws, we would expect to be able to show that orbital velocities will increase near a galaxy's nucleus and decrease at large distances from its center, just as occurs in our solar system.

But my Georgetown University team found that orbital velocities remain high at large distances from the galactic center. For some reason, our group didn't find it significant. No follow-up studies were made. I did not even stress it when I returned to the problem of galaxy rotation a decade later. Yet buried in the finding was an inescapable fact: The mass of a galaxy extends far beyond the stars, dust, and gas we can detect with conventional astronomical instruments. And although velocities of stars as they orbit their galaxy help us calculate the mass within that galaxy—making them a fundamental diagnostic tool for "weighing" not only galaxies but the entire universe—it was to be more than a decade before astronomers collectively recognized the need to measure star velocities across an entire galaxy disk.

APPOINTMENT AT THE CARNEGIE INSTITUTION

One day in January 1965, I had an appointment in a beautiful golden brick building set in a hilly, park-like setting in Washington, D.C. I was walking into the Department of Terrestrial Magnetism, familiarly known as DTM, one of the laboratories of the Carnegie Institution of Washington. I was there to ask for a job. This was not a sudden impulse. I had been working toward this moment for 13 years.

I first saw the DTM in 1952, when I was a graduate student in astronomy at Georgetown University. At that time, I had discussed the distribution of galaxies in the universe with George Gamow, a brilliant physicist/cosmologist and professor of physics at George Washington University. His questions ultimately became the topic of my doctoral thesis, and Gamow my thesis professor. With no common university, Gamow suggested that we meet at the lovely, cloistered DTM library. It was on my first visit to DTM that I decided that this was the place I would like to ultimately work. The research scientists there were attacking fundamental questions of broad interest, and were supported to engage in research of their choosing. I could not imagine a more inviting atmosphere in which to do research.

As a laboratory of the Carnegie Institution, the DTM is a sister institution to the Mount

Wilson Observatory in California. When I joined the staff of DTM in 1965, I gained access to a whole new way to observe faint objects. DTM was a leader in designing a new state-of-the-art device called an image tube spectrograph. My new colleague, Dr. Kent Ford, Jr., headed the project. The Carnegie Image Tube Committee, formed by Vannevar Bush in 1954, was a collaboration among astronomers, physicists, engineers, and industry. Its goal was to develop an electro-optical device that could amplify light by converting it into an electron beam focused by a magnetic field. The resulting image would be projected onto a special photographic phosphor plate. The idea was to make existing telescope/photographic plate combinations more light sensitive than ever before, and several teams were working on similar electronic optical systems around the world. Kent worked with engineers in the electron tube industry to develop the device, which he tested at DTM and installed at observatories worldwide in the mid-1960s. For a decade, until CCD detectors replaced the image tube/photographic plate combination, image tubes greatly extended the sensitivity of all telescopes, making those that were no longer competitive in regular photography and spectroscopy capable of contributing once again to the field of extragalactic astronomy.

Because of Kent's expertise and his well-designed spectrograph, he and I were early users of image tubes.

At a telescope, the spectrograph splits and spreads out light from a source—a star or galaxy—into its rainbow spectrum of colors. Hydrogen atoms within stars and gas produce very specific and easily recognizable spectral lines—and their appearance changes with the object's velocity toward or away from the telescope. An object's light shifts toward the red end of the spectrum when a source is moving away from the observer, and shifts toward the opposite, blue end when the source is moving toward the observer. Careful study of an object's redshift—or blueshift—tells not only which direction it is traveling, but also how fast. And measuring the velocities of faint objects from the shifted lines they produced was our goal.

April 1, 1965, the day I joined DTM, was also an exciting day for astronomers. That day's issue of the *Astrophysical Journal* contained an important letter by Maarten Schmidt announcing the spectra and velocities (redshifts) of five distant quasi-stellar objects known as quasars or QSOs. This brought the number of quasars with known redshifts to nine. These quasars appeared to be moving away from Earth at speeds that had never been observed in the universe before. Quasars were then, as now, enigmatic objects: extremely luminous, but with a source of energy that was yet unknown. In August 1965, just a few months after starting work at DTM, I joined Kent, who was spending the summer at Lowell Observatory in Flagstaff, Arizona, to observe with the 69-inch Perkin telescope. Because image tubes have a sensitivity extending far into the red spectral region compared to unaided photographic plates, Kent was observing faint red stars. I arrived in Arizona with astronomical finding charts of galaxies and quasars, interrupted his program, and we virtually never observed another star in our 25 years of observing together, shifting our attention to these new mysteries.

With the enhanced red sensitivity of the image tube we could detect the bright quasar emission lines that are beyond the spectral region available to photographic plates. After our very successful August observing session, we submitted a paper to the *Astrophysical*

SEARCHING FOR DARK MATTER

Astronomers have never come close to actually finding enough mass in the universe to produce enough gravity to counteract its expansion—to "close" the universe. The "gap" between the amount of mass detected and the amount needed has changed as new parts of the universe have been uncovered, or as the value of the Hubble constant, the acceleration parameter that counteracts gravity, has changed with new observations.

In the 1930s, Fritz Zwicky wondered what keeps clusters of galaxies together over time, given the high velocities they have within each cluster. Recently, astronomers have asked why tidal interactions between galaxies aren't more disruptive. And, why do the rotation curves of galaxies stay flat and not fall off with distance from the center?

The explosion of radio astronomy after World War II and the expansion of the accessible spectrum made possible by new telescopes changed how we look at the universe. Each time a new portion of the electromagnetic spectrum opened up to our telescopes, new components of the universe were detected. In the 1960s and '70s particle physicists were working on the fundamental structure of matter. In their atom-smashing accelerators, physicists discovered new families of particles with properties that could account for the bewildering phenomena that came about from these collisions at high energies. Were these new classes of particles far different from ordinary matter?

In the 1980s, the worlds of cosmology and particle physics found each other and the issue of dark matter came to the fore, as did a new way of thinking about the universe.

We know that dark matter interacts with ordinary matter through its gravity and, to a lesser extent, with the "weak force" that physicists postulate governs the decay rates of radioactive elements. Scientists have two categories of candidates for this dark matter: ordinary matter and exotic matter.

Ordinary matter candidates include massive compact halo objects, or "MACHOS," particles too dim to be seen at vast distances, such as brown dwarf stars in our galaxy, extra-galactic planets, and black holes. Studies show, however, that there is just not much material in this form.

The next ordinary matter candidate is the neutrino—low-mass but highly abundant chargeless particles that move at the speed of light. Neutrinos move so fast that their velocity would have to be slowed significantly to account for the dark matter. So far, there does not seem to be enough of them to close the gravity gap.

Now for something completely different: exotic mass. The first of these mysterious particles is called "AXIONS," hot particles with hardly a billionth the mass of an electron and thought to be vastly abundant in the halos of galaxies. Another candidate is "WIMPs," weakly interacting massive particles, thought to be cold particles 10 to 100 times more massive than protons. Physicists are searching for WIMPs in particle detectors, so far, with not much luck.

Every day scientists are turning up bits of data that could one day explain the source of that extra gravity we know exists but cannot see—so keep tuned!

—*David DeVorkin*

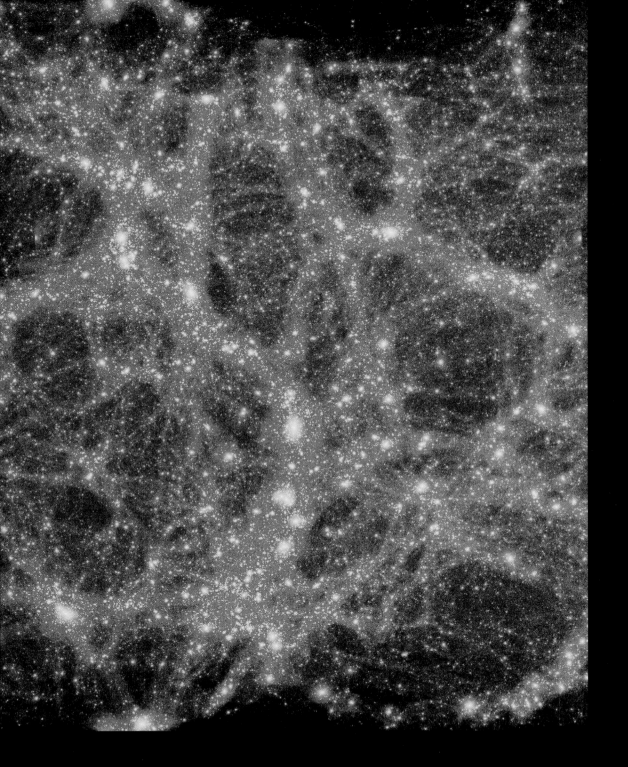

This simulation of cold dark matter shows how 80 million particles evolved over the age of the universe to the present epoch. Brighter colors represent higher matter density. Simulations such as this show how ordinary matter, in the forms of stars, galaxies, and clusters of galaxies organizes itself in the presence of dark matter.

WHY DOES AN OPTICAL ASTRONOMER STUDY SOMETHING SHE CANNOT SEE?

Journal in September that showed the spectra of three quasars. Although these quasars had known velocities, their red spectra had not previously been observed. We detected and measured significant amounts of H alpha—the strongest line of hydrogen. It was the beginning of a new understanding of quasars, which we now know are brilliantly bright nuclei of very distant galaxies powered by super-massive black holes at their cores. In the paper, we also showed the spectrum of a recently discovered infrared star, and a spectrum of the night sky extending far into the red. Because the objects we were observing were so faint, their radiation was no brighter than the radiation from atoms and molecules in the night sky. Hence the quasar spectra we obtained were "contaminated" with the spectrum of the night sky. To untangle the night sky radiation from the quasar spectrum, Kent and I also took and published a spectrum of the night sky alone, which revealed new information on what lies above our heads.

For the next two years, Kent and I observed quasars both at Lowell and at the new 84-inch telescope at Kitt Peak National Observatory, located southwest of Tucson on the Tohono O'odham (Desert People) Indian Reservation in the Sonoran Desert. Across this expanse, the

After measuring the rotational motions of spiral galaxies with the image tube spectrograph pictured above, Vera Rubin and W. Kent Ford came to the startling conclusion that stars do not slow down with distance from the center of a galaxy— proof that galaxies hold vast amounts of dark matter.

ancestral homeland of the Tohono O'odham Nation for more than 2,000 years, Kent and I ferried our 300-pound spectrograph, power supply, and numerous items of support equipment between Lowell and Kitt Peak. The Indian technicians who helped us unload the truck would remark, "There must be an easier way than this to earn a living." I was never sure whether they were talking about themselves or about us.

Our observations placed us in the exhilaratingly competitive world of quasar observers. Phone calls came from eminent astronomers to discuss spectra, and to exchange ideas, observing schedules, and plans. From those exciting days, two memorable events stand out. Cosmologists Bruce Partridge and Jim Peebles used our spectrum of the night sky in a paper titled "Are Young Galaxies Visible?" They calculated the probability of detecting highly redshifted young galaxies against the background of the night sky. I was enchanted that a spectrum of ours was involved in answering a question that I did not know enough to ask, although I had realized as I obtained the spectrum that this was totally new information for the astronomical community. I still believe that having astronomers use my data is the highest compliment I can receive.

The second event had a career-long effect. I concluded that I did not want to remain in the high-pressure field of quasar observing. Other quasar astronomers had significantly more access to big telescopes. They also had closer contacts to radio telescope observers, who enjoyed great success in discovering quasars because they stood out in the radio range of the spectrum. Kent and I had observed quasars because they were at the frontier of astronomy, and this was what astronomers did with a telescope capable of observing faint objects. But I also knew that we could profitably use the extraordinary sensitivity of the image tube to study regions in nearby galaxies that were formerly too faint for study. My education at a women's undergraduate college and at Georgetown University—both outside of mainstream astronomical education and research—had left me with a proclivity to ask unconventional questions that I thought would be valuable to pursue. I hoped to produce results that would be of great interest to other astronomers. I also hoped that I could work at a pace that would fit comfortably into my Washington life as a wife and the mother of four lively children.

By September 1967, Kent and I were back at Lowell and Kitt Peak on a long-term observing program to learn about the rotation of stars and gas in the Andromeda galaxy, more commonly referred to as M31 in astronomy circles. Our nearest large spiral neighbor, M31 is much like our own galaxy, and near enough so that individual stars and gas clouds surrounding the hot young stars can be identified. Observing individual stars was still impossible with available telescopes, but the hot gas surrounding the stars emitted much of their radiation in a few strong, bright lines of hydrogen, which we hoped could be measured.

As the stars and gas orbit the center of M31, those on one side of the disk are carried toward the observer and those on the other side away from us. By measuring the red and blue shifts of specific points within the galaxy with great accuracy, we deduced the velocity of almost 100 regions across the galaxy disk. Exposure times were typically one or two hours, but the regions were too faint to be seen in the telescope finder—a common problem when using the image tube to observe objects fainter than the telescope had originally been designed to observe. To compensate for this problem, before going to the telescope, we had to create what we call "offsets," calculated locations based on distance and direction to the nearest visible star. The details were exacting, and required that we make mistake-free calculations at the telescope, cold hands and dim red flashlights notwithstanding. The faintest regions required exposures of about six hours, so we could cover only two regions in a 12-hour night. As I guided those long exposures, seeing the faint glow of the M31 nucleus in the eyepiece, I often wondered if someone in M31 was observing us, and wished that we could exchange views.

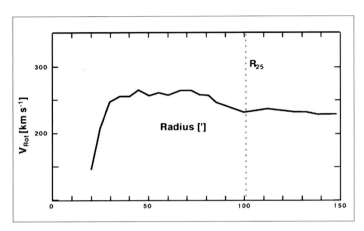

Plotting orbital velocities of stars in M31 from its center to its outer edge (l to r) produces this flat rotation curve, in which stars do not slow down with distance from the galaxy center.

For an observing run at the telescope, I usually arrived at the observatory a night or two early to get ready. One task was to use the darkroom to cut large photographic plates into two-by-two—inch squares to fit the camera in the spectrograph. This had to be done in total darkness. The emulsion side of the glass had to be carefully handled to ensure that no scratches or flecks of glass or dust remained because these could ruin the tiny details we would be trying to record. I then loaded the small plates into a rack and transferred them to an oven where they baked in dry nitrogen for 24 to 72 hours to enhance their sensitivity, a standard astronomical trick. We needed every photon of light that we could capture.

Just before the first observing night, the plates—now in a can filled with cool dry nitrogen—were transferred to a darkroom near the telescope. For each exposure during the night, I opened the can, loaded a plate into the holder, emulsion side toward the telescope light, refilled the can with dry nitrogen and closed it. Errors, such as turning on a flashlight or forgetting to close something, meant disaster. Generally, only the first exposure of a night was developed immediately upon completion—while the second exposure was gathering its photons. Remaining exposures were often not developed until morning unless there were two of us observing and our curiosity was too great to wait. One of us would guide the telescope, and the other would develop the little plate.

Today, with computer-controlled CCD detectors, computer screens display each telescope frame as it is read out following the exposure. That makes observing nights busier because frames come faster and there is always computer processing to do. But they are also less intense because an error is less devastating. Best of all, observing has become warmer because the observer sits at a computer console in an interior room. Still, I'm not sure which procedure is more fun.

A swirl of hundreds of billions of stars, M31, the Andromeda galaxy, resembles the
Milky Way. Rubin and Ford measured velocities of stars in the center and edges of
of the galaxy to determine that they were not slowing down with distance.

WHY DOES AN OPTICAL ASTRONOMER STUDY SOMETHING SHE CANNOT SEE?

INFLATION

There have always been some nagging problems with the big bang model. For about 30 years now, cosmologists have wondered why the universe looks much the same no matter where you look, even places separated from each other by distances greater than the travel time of the speed of light, over the lifetime of the universe (15 billion years). How do these distant horizons get connected? Equally puzzling, and recently set into sharp focus by new methods of measuring the distances to galaxies and by more detailed studies of the cosmic background radiation, is why the universe now seems to be so perfectly "flat." Dynamically, it seems to have just exactly the amount of material needed to prevent it from expanding indefinitely, and not enough to cause it to collapse. The universe seems to sit, miraculously, on a cosmological fence between infinite expansion and violent recollapse.

Flatness requires a very precise overall density of the universe and the density is an indicator of the amount of gravitating matter in the universe. Right now, flatness requires far more matter in the universe than has been so far detected. This gets us back to wondering about dark matter again and if there is sufficient dark matter in the universe to create a gravitational field sufficient to halt the expansion someday. But there is something else to consider.

About two decades ago, physicist Alan Guth found a way to account for the flatness problem and horizon problem by postulating that in the first instant of the very early universe, in its very first fraction of a second of existence, so small it cannot even be talked about among friends, the universe exploded at a rate far greater than anything seen or even comprehended today. It continued this wildly exponential "inflation" for another tiny fraction of a second until all of space-time within its volume was completely straightened out, or "flattened." Also, during this preamble to the universe as we know it, all parts of space-time were in contact with one another. In other words, an inflationary universe produces a flat universe, and a homogeneous one to boot.

Inflation remains controversial today, but it provides a surprisingly robust way to link together what physicists are learning about the early universe from studying reactions in atom-smashing particle accelerators with what astronomers are learning about the large-scale universe. One of the most fruitful recent studies has been the study of the spectrum of temperature fluctuations in the early universe by the COBE satellite and its successors on the ground and in space. As astronomers improve their view of the fine-scale structure of the cosmic microwave background radiation (see Wilkinson chapter) they have found that the temperature fluctuations are consistent with an extremely rapid and very short early expansion phase. This is not proof of inflation, but it is an indicator that it should be taken seriously.

This blend of particle physics and cosmology is at present the most popular and productive way to explore how the universe began and why it looks like it does today. But it requires some real mental gymnastics.

One of the most difficult aspects of inflation to accept is that it requires a very short

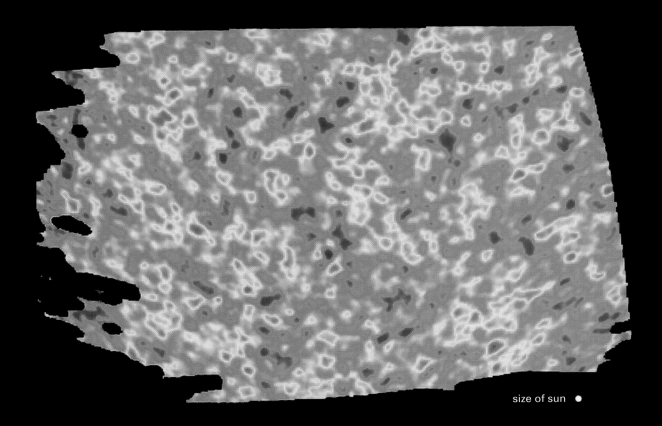

size of sun ●

period of time during which the universe expanded faster than the speed of light. This would seem to violate Einstein's teaching, but, in fact, what is expanding is not matter, but space-time. Matter is simply being carried with the expanding reference frame. What Einstein actually said was that nothing could move through space-time faster than light.

The concept of inflation has gone through many refinements since it was first suggested and discussed in 1980 and 1981 by Guth and others. Theorists like the idea because it allows them to consider the fascinating question of what came before, or even what produced the big bang. One form of inflation theory, called chaotic inflation, allows theorists to think about how our universe could have sheared off as a quantum fluctuation from some preexisting continuum. Universe might therefore breed universes.

—*David DeVorkin*

A picture of the early universe, this BOOMERANG image shows minuscule temperature variations comparable to the apparent size of the sun in the sky (lower right corner of image) in the faint afterglow of the big bang.These ripples in the primordial plasma have since evolved into clusters and superclusters of galaxies. BOOMERANG's results reveal that the size distribution of the structural features seen in this small section of the sky are just what cosmologists would expect from a universe whose geometrical curvature was infinite, or "flat."

WHY DOES AN OPTICAL ASTRONOMER STUDY SOMETHING SHE CANNOT SEE?

From the spectra of M31 I measured velocities and plotted them on a graph representing the entire visible galaxy, from its center to its outer limits. As we had discovered back at Georgetown University, my resulting curve showed that rotation velocities remained high even at the outer edges of the optical disk. The distribution of light in M31 was not a clue to its distribution of mass. Again, unlike the planets in our solar system, the motion of the stars in M31 did not slow down with distance from the center. And again, as we had done in the 1960s, we stressed the observation and chose not to speculate about the galaxy's mass beyond its optical limits. Only looking back some time later did we realize we had uncovered convincing evidence of a form of matter that was invisible to our eyes and instruments, yet by far the most dominant feature of the universe: dark matter.

A DARK UNIVERSE

By the mid-1970s, evidence was accumulating that a large portion of the mass in the universe might be dark. By "dark" astronomers mean either unseen or technically unobservable by present means. Some believe that this dark matter is regular matter—composed of protons, neutrons, and electrons just like those that make up our world and ourselves. Others believe that it is a wholly new and unknown form of matter, or something that manifests the presence of mass, such as a gravitational field, but that otherwise is not like the matter we know of and can perceive. One way or another, to really understand the distribution of mass in galaxies, and in turn the distribution of matter in the universe, Kent and I recognized that a collection of rotation curves for a wide variety of spirals would be required. This observing program had to await the completion of the new Kitt Peak four-meter Mayall telescope, for which Kent helped to build an even larger spectrograph.

With the opening of the twin four-meter telescopes at Kitt Peak and at the Cerro Tololo Inter-American Observatory in Chile, astronomers had a powerful new set of telescopes available on an open and competitive basis, irrespective of whether their home institutions owned a part of it. We applied for telescope time, were accepted, and started observations for spirals with well-defined but different structures.

For these observations, we chose galaxies that appeared small enough on the sky to fit completely on the long slit of the spectrograph. Rather than observe individual emission regions as we had done for the nearby Andromeda galaxy, now we made a single exposure on a slice completely across the disk of the galaxy. We were still restricted to observing the gas; stars are too faint even for this generation of larger telescopes.

Those nights at the new Mayall telescope were among the most wonderful nights of my observing career. Before the slit-viewing optics were complete, we had to set each galaxy on the spectrograph slit visually, standing and looking through the telescope. Those views of galaxies, through a telescope with a mirror of more than 150 inches in diameter, were the grandest galaxy views I have ever had. The observations too were exhilarating.

On a single night, we could obtain four spectra of more than two hours each. The first night, our curiosity was so great that I developed each plate as it was completed. My first reaction was one of joy, just to see that we really had obtained emission lines all across the galaxy, even from the faintest outer parts. I next realized that all the spectra looked alike, blue-shifted emission lines on the side of the galaxy that was approaching us, a normal spectrum at the bright, tiny nucleus; and red-shifted emission lines showing a consistent velocity all the way out the side of the galalxy receding from us.

By 1976 we knew that rotation velocities of a few very luminous galaxies remained flat across the galaxy disk—contrary to the diminishing velocity scientists would previously have expected to find. By 1977 we could publish spectra of ten galaxies, and a reader could see the data with his or her own eyes. Few astronomers were nonbelievers after seeing these images. Within five or six years, we obtained high-accuracy rotation measurements of more than 60 spirals, distributed evenly among the structural types (based on the relative size of the galaxy's central nucleus and the tightness of its spiral arms). Although they looked very different in optical wavelengths, all showed flat constant-velocity rotation curves.

This is what we learned about galaxy masses: We do not know the total mass of a single galaxy; we know only the mass out as far as we can find moving things to measure. For galaxies that have disks of neutral hydrogen that extend beyond the optical galaxy, radio observations reveal that rotation velocities remain high across these gas disks also. The hydrogen gas is not the dark matter, but it too responds gravitationally to the presence of dark matter. Astronomers call the distribution of dark matter a halo, but actually it is a sphere-shaped distribution surrounding the galaxy disk. Estimates indicate that at least 90 percent of the mass of a galaxy is contained in unseen dark matter, although there is currently no way to directly detect it.

As Margaret Geller points out in her chapter, Fritz Zwicky and Sinclair Smith noted in the 1930s that the high relative velocities of galaxies in clusters suggested that these clusters should be dissolving, which they were not. They concluded that there must be nonluminous matter in clusters that holds these clusters together. Our velocity studies helped to establish that dark matter is a necessary and major component in single galaxies as well.

Why did it take about 40 years for the conclusion of Zwicky and Smith—that clusters of galaxies contained unseen matter—to make it to mainstream astronomy? For one thing, some astronomers did not believe the cluster results, arguing that perhaps clusters were actually dissolving. Not so. For the majority of astronomers, the results were not actively disbelieved; they were more of a curiosity, not fitting into the fabric of astronomy at that time.

However, by the mid-1970s, there were other reasons to believe that dark matter existed. Theoretical studies of disk galaxies by Jeremiah Ostriker and Jim Peebles at Princeton in the early 1970s had suggested that disks were grossly unstable and would eventually form a bar, or puff up, but that they could be stabilized by a halo whose mass was larger than that of the disk. In 1974, Ostriker, Peebles, and Amos Yahil introduced a paper in the *Astrophysical Journal* with the words, "There are reasons, increasing in number and quality, to believe that the masses of ordinary galaxies may have been underestimated by a factor of 10 or more." Reasons continued to accumulate throughout the decade.

Studying something that you cannot see is difficult, but not impossible. Not surprisingly, astronomers currently study dark matter, which we cannot see, by its effects on the bright matter that we can see. We can deduce a few characteristics of this matter: It is less concentrated at a galaxy's center than is light, it extends well beyond the optical boundaries of a galaxy (our halo might even brush the halo of M31), and its form is not as flat as a disk. Clumps of dark matter that formed in the early universe may have been the regions in which galaxies would later be formed by infalling matter.

The amount of luminous matter in the universe is small. It accounts for less than 1 percent of the matter whose gravity would be sufficient to halt the expansion of the universe. Even the amount of dark matter needed to account for the galaxy and cluster velocities that we observe still equals only about 20 or 30 percent of the matter needed to halt the expansion of the universe. This value is larger than the amount of "normal" matter produced in the early universe, according to cosmologists, so some of the dark matter must consist of exotic particles permitted by theory but not observed to date. That could change, though, as particle scientists design clever searches to find them in large accelerators.

There are other ways to infer invisible matter. One exciting effort is to detect Einstein rings—gravitationally distorted multiple images of a distant background galaxy (or galaxies) hidden behind massive galaxy clusters. Nature made telescopes before Galileo did!

Those of us whose studies have contributed to the need for dark matter are aware of one caveat: Dark matter is necessary because Newtonian gravitational theory, coupled with the luminosity distribution in galaxies, predicts that rotation velocities should fall for regions well beyond the nucleus. They do not. Early in the 20th century, physicists learned that, in domains as small as atoms and nuclei, this conventional rule of physics was not valid. We have only recently been testing these laws on scales larger than the solar system. Here, too, they have failed, but we attribute their failure to the existence of dark matter. Not until we have identified the dark matter will we be free from the very slight but non-negligible possibility that someday our students will enthusiastically embrace entirely new rules of physics for galaxy-scale systems.

We live in an enormously complex universe. There are many unknowns in it. We cannot feel overly confident that we comprehend major features of our universe when we do not know the composition of at least 90 percent of its mass. Because the amount of matter in the universe is fundamental to understanding whether the universe will expand forever or will ultimately halt, the study will continue. There will be no shortage of subjects for our grandchildren and their grandchildren to study. Perhaps the fundamental question is, Will we ever be able to know it all?

The distinctive Sombrero galaxy may, like other galaxies, reside within a halo of unseen dark matter that extends well beyond its rim. Or could the gravitational effects attributed to dark matter result from an incomplete understanding of how physical laws work on a cosmic scale?

WHY DOES AN OPTICAL ASTRONOMER STUDY SOMETHING SHE CANNOT SEE?

THREE-DIMENSIONAL MAPPING OF THE DARK UNIVERSE

J. Anthony Tyson

The sun has set at Cerro Tololo and the last glimmers of the red sky fade to darkness. Ahead looms the great dome of the observatory on this arid, dusty saddle high in the Chilean Andes. I look into the night sky, into the darkness of the great rift between the star clouds of Sagittarius, and recall a story about how Australian aborigines' cosmology centers on the dark "lagoons" between the stars. "Did they get it right after all?" I wonder.

In remote places like this, the panorama of the night sky unfolds as a scroll of darkness punctuated by light. The vast areas between the stars and galaxies appear empty and dark. Naturally drawn to the light, astronomers once passed up these dark places. Much of our knowlegde of the universe has come from distant sources of radiation: infrared, microwave, optical, and

In this illustration of gravity's effect on space and time, a massive complex of invisible "dark matter"—color coded red—warps the fabric of space-time for millions of light years around. Gravity affects the way matter and light move.

x-ray. Modern cosmology has been built on two pillars of radiation: the residue from the big bang and the distribution and spectra of stars and galaxies.

However, direct study of the luminous universe tells only part of the story. It has long been known that mass, not luminosity, is the key to the structure of the universe. This is because gravity plays a central role in how things move. Structure occurs in the universe because mass is inherently unstable gravitationally. Over cosmic time, due to gravity, "over-dense" regions become still more dense. Tiny ripples of density existing 300,000 years after the big bang have grown into the complexity of mass structure—galaxies to superclusters of galaxies—we now see some 14 billion years later. On the largest of scales, the overall expansion history of the universe is governed by its mass-energy (Einstein taught us that mass and energy are related). But because mass could not be seen directly, we have long had to use luminosity to tell us about its invisible brother, mass.

Over the years, we have given names to the particles that make up the mass that produces radiation, from radio through optical to x-ray and gamma-ray radiation. These particles—electrons, atoms, and nuclei—are the building materials of our "familiar" visible universe. Now, however, after more than a two-decade harvest of observing, testing, and debating, it seems that this ordinary matter, the stuff we are made of, cannot be the chief component of most of the mass in the universe. Instead, we now know that more than 96 percent of the universe is dominated by unknown forms of mass—we call it dark matter—and energy, also known as dark energy. A huge amount of dark matter—roughly ten times as much mass as there is in all the stars and gas and dust combined—controls the early evolution of structure in the universe. The dark matter is thought to be some very different kind of particle created during the hot big bang, and it interacts with our familiar particles only weakly. Underground detectors of these "weakly interacting massive particles" are being developed. So far, though, we have no evidence for them aside from what we've seen: massive gravitational effects that indicate clumps of dark matter larger than the largest visible galaxies.

COSMIC MIRAGES

So how do we study dark matter if we cannot see it? The internal dynamics of galaxies (see Vera Rubin's chapter) can tell us about the distribution of all forms of mass in the luminous parts of galaxies. But we now know that galaxies are surrounded by a halo of dark mass. On these larger scales—massive halos of galaxies, clusters of galaxies, superclusters, and massive dark structures spanning billions of light years—we rely on the gravitational influence of these dark matter structures themselves. Their mass will bend light rays from background galaxies like a lens and create what astronomers refer to as "cosmic mirages."

Scientists have speculated about the light-warping properties of mass for more than two centuries. Albert Einstein's general theory of relativity predicted the exact degree to which the mass of the sun could bend starlight. Solar eclipse expeditions in 1919 confirmed Einstein's prediction, and became the key observational tests of his theory. This success stimulated others to think about the ability of foreground stars to bend the light of background stars, a phenomenon we call "gravitational lensing." In 1937, Caltech astronomer Fritz Zwicky proposed that gravitational lensing of a background galaxy by a foreground galaxy's mass was a "near certainty." In studies of the motion of clusters of galaxies, Zwicky had found that galaxies were far more massive than had been generally thought, but his gravitational lensing idea was largely ignored and lay dormant for nearly half a century. Now, however, with far larger telescopes and more powerful imaging technologies available, there has been a rebirth and exploitation of gravitational lensing as a tool for studying the cosmos.

In a curious way, the universe is connected on its largest and smallest scales. In the earliest moments of the universe, tiny dark matter particles were created amongst temperatures and energies far higher than any imaginable on Earth. Today, that legacy is detectable in its cumulative gravitational effects on large-scale structures in the universe. The worlds of particle physics and astronomy are coming together in a transformed worldview. Copernicus displaced Earth from a central position, and Harlow Shapley and Edwin Hubble removed our galaxy from any special location in space. Now, even the notion that the galaxies and stars comprise most of our universe is being abandoned. Emerging is a universe largely governed by particles of dark matter and, we are beginning to think, by an even stranger dominance of a smoothly distributed and pervasive dark energy. The aborigines of Australia had been right.

How can cosmic mirages reveal "images" of dark matter? Imagine a room covered— walls, ceiling, and floor—with wallpaper. The wallpaper is decorated with a reliably repeating pattern of very close dots. Now imagine a magnifying glass suspended between you and one wall, a glass so clean and flawless the glass itself is invisible to you. Yet as you look in its direction, you'll notice that the pattern of dots behind it is distorted, as the light bends through the glass. By observing the pattern of distortion, you can get a very good idea of the size and shape of the magnifying glass, even though the glass itself is invisible to you. Similarly, the light bent by a clump of dark matter can help astronomers determine the shape of that clump, and map it.

To search for the cosmic wallpaper, we have to look deep enough into the background universe so that there are thousands of galaxies projected near the foreground lens. And that begins with what astronomers have long known: That if you build a bigger telescope and develop faster, more sensitive detectors of light, you will see deeper into the universe. There were always more galaxies to be seen, and tricks to be learned to see them. Five decades ago, photographic emulsions typically recorded only about one quantum of light for every 100 hitting them. By the mid-1970s, engineers at such places as Eastman Kodak, in collaboration with astronomers, had pushed the sensitivity of photographic plates to record-high levels, so that at least one of every 20 photons hitting the plate was recorded. Still only 5 percent efficient, this was a great boost. In fact, it helped Richard Kron and me find the first hints of the cosmic wallpaper in the late 1970s.

As is often the case in research, the discovery of the wallpaper came as a by-product of unrelated projects. Working as a physicist at Bell Labs, I was naturally interested in pushing technology. I was also attracted to problems associated with seeing the faintest objects in the

GRAVITATIONAL LENSING

The lenses most people are familiar with are made of glass, but anything that can alter the direction of a light beam can act like a lens. Air pockets of different temperatures can bend light to produce a mirage on Earth, and, as Einstein predicted in his relativity theories, so can strong gravitational fields.

In the diagram here, light rays from a distant blue galaxy are bent by the gravity of an intervening clump of matter marked by a cluster of bright galaxies. These galaxies are only a very small part of the entire mass of the cluster, which is made up predominantly of dark matter evenly distributed between the bright galaxies. Together, they form a smooth-acting lens to create a gravitational mirage: The galaxy located behind the "lens" appears at different places on the sky and is distorted into a series of arcs or rings of light when viewed from Earth. The light rays are bent by gravity, and an observer on Earth will see them in this distorted form. The mirage is the ring that extends from the observer's eye through the edge of the cluster of galaxies and out into indeterminate extragalactic space. From Earth, the ring simply appears projected onto the sky.

Astronomers now realize, from the growing presence of these gravitational mirages as detected by Earth's largest telescopes, as well as the Hubble Space Telescope, that mountains of dark matter exist throughout the universe, altering the observed distribution of matter. Although we don't know where the distant blue galaxy would have appeared on the sky in the absence of the dark matter responsible for the lensing, we can measure its shape. The sky is covered with billions of distant galaxies and their telltale systematic distortions reveal the presence of foreground dark matter. These coherent distortions of galaxies in any part of the sky can thus be transformed into a map of the foreground dark matter and to a new image of the "dark" universe.

—*David DeVorkin*

Light from a galaxy lying directly behind a galaxy cluster is split into a ring of multiple warped images—a gravitational mirage. In rare cases when a distant galaxy is by chance projected exactly behind a massive clump of mass (shown here associated with a cluster of galaxies), multiple magnified images of this background galaxy may be formed. Shown here are three possible paths for the light, corresponding to three images of the background blue galaxy. Blue arcs around some clusters were discovered independently in 1986 by U.S. and French astronomers. Such dense clumps of dark matter are rare. More often the positions of distant galaxies are only slightly perturbed by intervening mass, and no multiple images are formed.

THREE-DIMENSIONAL MAPPING OF THE DARK UNIVERSE

universe. I knew that so-called "radiogalaxies" have microwave "hot spots" or "lobes" on either side of the parent galaxy, and that one peculiarity of this radio energy is that it is not caused by a thermal source; it is caused by electrons accelerated to very high velocities. Radiogalaxies were known to emit "synchrotron radiation," at radio wavelengths when energetic electrons slammed into magnetic fields. But what caused the acceleration of the electrons in the first place?

To find out, Phil Crane, Bill Saslaw, and I applied for telescope time at the Kitt Peak National Observatory in Arizona, and eventually secured several observing runs on the Mayall four-meter reflector, the largest telescope at the observatory. We started the project by mounting a highly sensitive photomultiplier, an instrument used to detect and amplify light from faint sources, at the focus of the telescope, and scanned across the places where the radio lobes were. There were problems with using this technique because the instrument's photoelectric system and the Earth's atmosphere both changed from minute to minute. Occasionally, we saw evidence of a faint glow from the radio lobes, but there were too many other effects that we could not control. So we turned back to photography, but this time with a twist. First, with the help of Art Hoag at Kitt Peak, I learned how to "push" the efficiency of special photographic emulsions to the limit, and exposed

210

Invented in the early 1970s, charge-coupled devices, or CCDs, have revolutionized astronomical imaging. Wafer-thin silicon chips (center) that are sensitive to light, CCDs have replaced photographic plates.

them at the focus of the Mayall telescope for one hour. I took multiple exposures of several of the radio galaxies, each time using a special eight-by-ten-inch Kodak plate. These plates were so sensitive that they had to be developed with a custom process at the end of each night. Of course, each plate covered an area on the sky far larger than we required for our project—and this turned out to be important.

Next, we converted the images on the plates to digital data by scanning each plate with a light beam. We also used special software for faint light sources to bring out detail. This process produced "noisy" images peppered with false "galaxies" caused by chemical impurities in the emulsion. Astronomers at that time believed this effect set the faintness limit for galaxy detection. But we wanted to go fainter. Could we? And if so, how? Arizona's version of a monsoon provided an opportunity to find out.

With our project half finished—with only a hint of light from those elusive radio lobes—I flew back to Tucson for another multiple-night run on the four-meter telescope. Then the rainstorm hit. Unable to use the telescope on Kitt Peak, I spent the week in the basement of

the headquarters building in Tucson with our collection of plates. Frustrated by the lack of progress and with nothing else to occupy my time, I decided to scrutinize the "false" galaxies on various plates covering the same region of the sky. Most were clearly chemical events in the emulsion, accompanied by occasional pieces of dust and lint that did not appear on more than one plate. However, some of the fainter dots were present on multiple plates covering the same area of the sky. These barely perceptible black smudges on the digitally stretched negatives appeared at exactly the same sky coordinates each time. I first tried counting these "common" smudges, but soon realized that my learning curve was biasing the counts. In an attempt to get to even fainter light levels, I used a simple program to digitize images covering a common portion of the sky, and then added them together, generating a deeper image. At its faint limit, this deeper image revealed even more.

The detail a CCD can resolve depends on the number and size of its tiny silicon light detectors, called pixels. Improvements in silicon chip production have made possible increasingly larger CCDs with higher pixel densities. The largest (bottom) is now used in multiple arrays to detect lensing across large patches of sky.

THREE-DIMENSIONAL MAPPING OF THE DARK UNIVERSE

Obviously I was seeing out to great distances through a forest of galaxies spread over a huge volume of the universe. There were enough galaxies in every little patch of sky (about 6,000 in an area the size of the full moon) that it seemed possible that, within any narrow view of a foreground galaxy, you'd also get chance projected images of galaxies beyond them.

I was so absorbed in thought I was nearly hit by a car while walking back from the Kitt Peak Headquarters on the University of Arizona's Tucson campus. "What if each foreground galaxy had a large mass?" I wondered. Light rays from background galaxies projected within a few arcseconds would be bent by this mass. Each background galaxy would be moved to a new place on the sky, systematically warping its image. This might lead to a new and powerful means to statistically weigh the mass of a galaxy!

It would take two decades and the development of a new imaging technology to fully exploit the cosmic mirage of distant galaxies, but later in the 1970s, Richard Kron discovered a hint of things to come. While studying the faint galaxies, he found that the galaxies at the faint limit of photographic plates were systematically bluer than brighter galaxies. The fainter and more numerous blue galaxies would eventually become useful tools for exploring mass.

Gravity from a continuous field of dark matter produced this ring of distorted "ghost galaxies" around galaxy cluster CL0024, 4.5 billion light years away. In reality there is only a single blue galaxy; it lies behind the cluster some 10 billion years away.

GRAVITY WARPING

By 1977 the idea of gravitational lensing was very much in the air in scientific circles. Unaware at the time of Zwicky's ideas, I recalled a 1973 article by Bill Press and Jim Gunn in which they asked the question, "What would the sky look like if every galaxy had so much mass that their gravitational attraction would eventually halt the expansion of the universe? Although the real universe turned out to have much less mass—and much less of it concentrated in individual galaxies—the notion of how foreground clumps of mass affect images of distant galaxies was shown in one of their figures. It would be two years before Doug Walsh, Bob Carswell, and Ray Weymann discovered the first gravitationally lensed quasar, but when I saw the Press and Gunn diagram I was immediately struck by the possibility of statistically estimating the mass of foreground galaxies using deep wide-field imaging. Any small image warp due to lensing would not be detectable in an individual background galaxy because we do not know its intrinsic shape, but analyzing tens of thousands of foreground-background pairs could bring the signature of this systematic lens warp out of the "noise" of random galaxy shape variations.

How could one possibly count and measure the brightness and shapes of thousands of galaxies? Several things would be required: wide-field deep imaging covering many square degrees of sky, control of image shape errors due to aberrations in the telescope and camera optics, and software for analysis of the shapes of thousands of faint galaxy images. The time had come to automate the detection and measurement of all objects on these deep images of the universe. My first step was to recruit someone from a group at Bell Labs developing software for pattern recognition. My collaborator was John Jarvis, an expert in pattern recognition and an amateur astronomer. Over the next three years we

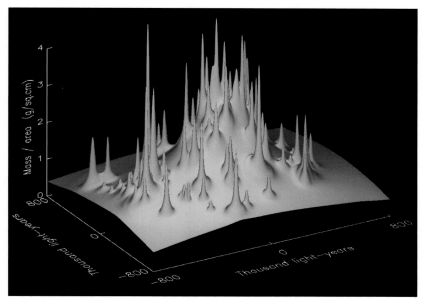

213

This graphic reveals the distribution of dark matter in the galaxy cluster on the opposite page. Spikes indicate some dark matter clings to single galaxies; most is spread smoothly throughout the cluster. Placing a grid behind this "lens" produces the warped grid on page 205.

developed software to automatically catalog the brightness and shape of thousands of faint galaxies. By early 1980 we had a glimmer of success. Examining the distortions of tens of thousands of faint galaxies appearing on the sky near brighter galaxies, we found a slight tendency for the faint galaxies to be distorted in exactly the way predicted—by lensing due to foreground galaxy mass.

One pressing scientific question was (and still is), How large and massive is the average galaxy's halo—the sphere-like cocoon of dark matter that surrounds it? Faced with our marginal detection, we decided to publish an upper limit to the effective mass and dark matter halo around galaxies. Of greater cosmological significance is the warp, or shear, of the distant universe, caused either by rotation of the universe or by vast overdensities of mass in the foreground. We searched for this cosmic shear also, but we found nothing on our first try. Photographic plates were just too unstable. Each exposure was on a different plate with its own peculiarities. Fifteen years later, we would achieve a solid detection of cosmic shear, thanks to a new imager called the "charge-coupled device," or CCD.

The origin of today's pervasive electronic imager—from the familiar digital snapshot camera to the hundred-megapixel wide-field imagers used in astronomy—was a device invented in 1970 for the purpose of storing an audio message. Within hours of hearing of the need for a solid-state scrolling memory, George Smith and Willard Boyle at Bell Labs invented the charge-coupled device using silicon-integrated circuit technology. The first CCD they made was a single one-dimensional "bucket brigade" for charged particles. It worked, but had an annoying sensitivity to light! The rest is history.

The first CCDs were not reliable imagers. However, by 1978 several of us were building experimental low-light-level cameras using small CCDs. This smallness was the one disadvantage of early CCDs; they covered 4,000 times less area than photographic plates, which were still being used for surveying the sky.

A CCD image from the Hubble Space Telescope of galaxy cluster Abell 1689, showing evidence of both strong lensing (multiple images of the same galaxy) and weak lensing (warped images of individual galaxies).

214

But CCDs offered three huge advantages over photographic plates: They were 50 times more sensitive to light, they were not subject to the distortions that can occur in the wet photographic developing process, and the same detector could be used for each exposure. This last advantage turned out to be most significant. After a year of experimenting with exposure strategies and image analysis software, we developed the "shift-and-stare" imaging technique. Multiple, slightly shifted exposures of the same area of the sky produced all the data required for filtering out the imperfections on the CCD and producing a spectacular deep image of the sky.

In 1983 Patrick Seitzer and I placed a small CCD at the focus of the four-meter telescope at Cerro Tololo Inter-American Observatory in northern Chile. Making multiple exposures through three colored filters, and using the "shift-and-stare" technique, we produced a stunning image of a small patch of the sky showing more than 400 faint blue galaxies. A deep photographic plate I had taken earlier showed only two stars in this same patch of sky! By simple extrapolation there had to be tens of billions of these faint blue galaxies over the sky, a perfect candidate for the cosmic wallpaper. We had discovered a vast and deep sea of faint blue galaxies extending out billions of light years.

It would be years before CCDs were large enough to be useful for wide-area surveys. Meanwhile, however, even the earliest CCDs provided a spectacular new view of small-scale lensing in strong gravitational fields.

Although the Hubble Space Telescope's (HST's) field of view is small, the four 800-by-800-pixel CCDs in its wide-field planetary camera have generated an unprecedented high-resolution view of our universe. What do we see when we point the HST at a place on the sky where there is a known supermassive gravitational lens? As light travels to us from a distant source, its rays are deflected by the gravity of intervening masses. The deflection is noticeable only when the inter-

215

If dark matter were clumped around individual cluster galaxies, the image of Abell 1689 at left would resemble the simulation above. Mirages surrounding each galaxy do not appear in the actual image, a clue that dark matter is distributed throughout the cluster.

vening mass is huge and so concentrated that it forms multiple, highly magnified images of a background source (so-called strong lensing). The first strong lensing source was discovered in 1979 when a distant quasi-stellar radio source (quasar, or QSO) was shown to be lensed by a foreground galaxy, appearing as two point-like images. Since then about 20 more examples of multiple-lensed quasars have been found, some of them displaying four images of the distant source. Cosmologists have used the statistics of lensed quasars to constrain the dark energy content of the universe and as an independent measure of the expansion of the universe. Even with the resolution of the HST, a quasar's image appears as an unresolved point of light. In other words, it has no detectable dimension and is seen simply as multiple points of light on the sky. An important clue about the lens—the warped shape of the source—is thus missing. Because we cannot see the warped shapes of these images it is difficult to learn much about the distribution of mass in the lens. This mass distribution could tell us about the nature of dark matter. So what we need are sources whose shapes are visible. In other words, extended, resolvable objects like galaxies are required. And so is luck. That distant source galaxy must be directly behind the lens. If we could find an example of a gravitational lens producing multiple images of the same background galaxy, we could learn a lot about the detailed mass distribution in the "lens."

A cluster of galaxies makes an easily identifiable gravitational lens. Its mass bends light rays from other galaxies, producing a strong warp in their images. In the late 1980s, while looking for a background galaxy whose image was warped by an intervening cluster, we got lucky. David Koo found several faint blue arcs around a cluster called CL0024, and I took deep CCD images of it. They looked like multiple-lensed images of a single galaxy, but we needed to sharpen the image to be sure. Edwin Turner and Wes Colley of Princeton and I wrote a proposal to the Space Telescope Science Institute in Baltimore. When NASA pointed the Hubble Space Telescope at CL0024 it revealed five warped and highly magnified images of the same distant blue galaxy. Looking at the HST raw images for the first time we could see immediately that the same oddly shaped blue galaxy appeared many times! The massive gravitational lens, together with HST, had created a highly magnified view of a small patch of the distant universe that by chance contained a galaxy.

More than 150,000 faint blue galaxies—most more than seven billion light years distant—emerge in this deep sky image created with a mosaic of CCDs.

The next step was harder. Would it be possible to determine the detailed mass distribution in the cluster CL0024? The multiple images contain clues to the mass of the lens and its distribution because these images are formed by different rays from the background galaxy passing through the gravitational lens at multiple locations. At Bell Labs, Greg Kochanski, Ian Dell'Antonio, and I decided to explore all possible arrangements of mass in the lens that could result in the complex mirage we were seeing. This was a huge computational challenge. We had 10,000 mathematical constraints from the color image data, and a lens mass model with 500 adjustable parameters!

We developed a software optimizer for the solution, and after a year of supercomputer time we had a robust map of the mass in high resolution. It showed that while less than 15 percent of the dark matter sticks to galaxies in a cluster, most of it is diffusely distributed. These "soft mass cores" contradict current computer simulations of "cold dark matter" cosmology.

Soft mass cores have now been seen in high-resolution mass maps of two galaxy clusters. It is possible that new physics, not included in the "cold dark matter" computer simulations, will be required to explain it. While the dark matter particles interact with ordinary matter only weakly, might they interact strongly with each other? This could explain the soft dark matter cores in clusters, but there may be other explanations. After years of study we still do not know what makes up the dark matter, but observations of the way it clumps can rule out theories of what it is.

One way we can study dark matter densities is through what we call "statistical weak lensing." This new field of astronomy, which involves imaging mass instead of light, emerged as a result of our detection of faint and distant blue galaxies, the development of large CCDs and huge new optical telescopes, and improvements in microelectronics. Statistical weak lensing uses the telltale warp of all the thousands of background galaxies to derive a map of the foreground mass. No special alignment of a single source galaxy is required, and there are more than 500,000 faint background galaxies in every square degree of sky. Several groups of astronomers are now making maps of the dark mass distributions in the regions around massive clusters of galaxies. Here the huge mass associated with the cluster produces the most noticeable warps.

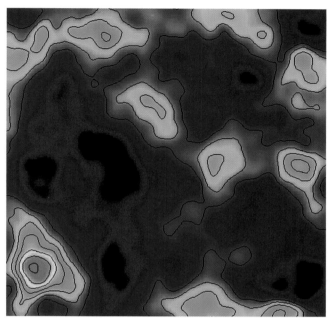

The distribution of mass in the image on page 216 revealed in 3-D mass tomography: A cluster of dark matter is the red peak in the bottom left corner.

How much dark matter in our universe resides outside clusters of galaxies? If most of the mass in clusters is in a smooth distribution extending out millions of light years, perhaps most of the dark matter of the universe is distributed more broadly than clusters. Think of clusters as the Mount Everest of the universe. Everest is huge, but most of the mass in Earth's mountains is in the more numerous smaller foothills. Weak gravitational lensing offers the potential to make mass maps of any part of the sky. We can "prospect" for piles of dark matter, big and small. Though we are currently limited to fields of view less than a degree across, telescopes are being used to probe this universal dark matter distribution. One such probe is a technique called "cosmic shear." Light rays from very distant galaxies pass by multiple clumps of dark matter on the way to our telescope. The cumulative effect of these near misses leaves a signature on the warp pattern. Depending on how the dark matter is distributed—in rare big clumps, or perhaps many more small clumps—the statistical pattern of warps of the distant galaxies will be different. By 1997 we were using our new CCD mosaic camera in the search for cosmic shear. Two years later we had a solid detection, ruling out the then "standard" dark matter model. Two other teams have obtained similar results.

A UNIVERSE OF MASS

Let's now have a quick look at this new universe of mass. Our new observational maps cover a range of scales and can constrain the nature of dark matter. There are big and small clusters of mass, and there are what appear to be filaments of mass. There is, in fact, enough mass to add up to about a third of the density that, in the absence of dark energy, would be required to ultimately slow the expansion of the universe to zero. We have found a complex universe of mass: dilute filaments of mass coexisting with piles of dark matter centered on clusters of galaxies. This complex dark matter structure took billions of years to grow. Its growth rate is predictable for a given model of the expanding universe. Probes of other features of the universe—from primeval deuterium to tiny fluctuations in the heat left over from the big bang to supernovae at large distances—suggest that some form of dark energy, when combined with the gravity of the dark matter, creates a flat cosmic geometry.

To determine what dark energy is and how we can probe its physics, we look indirectly at its influence on the expansion rate of the universe. Dark energy acts against gravity, tending to accelerate the expansion. Measuring the rate of structure growth by taking "snapshots" of mass at different cosmic times would provide clues to the nature of dark energy. In turn, this would tell us something about physics at the earliest moments of our universe, setting the course for its future evolution.

The world of quantum gravity at a fraction of a second after the big bang—when the universe was so hot and dense that even protons and neutrons were broken up into a hot soup of quarks—connects to today's vast expanding cosmos extending out 14 billion light years. Dark energy and dark matter are relics of the first moments when unfamiliar physics of quantum gravity ruled. A route to understanding dark matter and probing the nature of dark energy is to watch mass structures forming over the last half of the age of the universe.

So far, we have used the faint blue galaxies only as two-dimensional wallpaper. But what if we could gauge the distances to the billions of faint blue galaxies, placing them in four-dimensional space-time? The faintest galaxies have a range of colors, each one's color depending on its type and distance from us. The most distant galaxies have their spectra shifted to longer (red) wavelengths, and their light has taken up to ten billion years to travel to us. Using the colors of the galaxies, it is possible to gauge their distance. The more distant the source, the more warped its image. This is the clue that unlocks the universe of mass in three dimensions. If there is a foreground mass, the mirage effect on the background galaxies gets stronger for more distant galaxies. By measuring both the warp and the distances to the background galaxies, it is possible to reconstruct the mass as before and also to place the mass at its correct distance. Using 3-D mass tomography, David Wittman and a group of us recently detected, weighed, and placed at its proper distance a clump of dark matter associated with a small cluster of galaxies. This enables the exploration of mass in the universe, independent of light, because only the light from the background galaxies is used.

Mass tomography in 3-D opens a new window on the time evolution of the universe. By exploring mass in the universe in 3-D we are also exploring mass at various cosmic ages. This is because mass seen at great distance is mass seen at a much earlier time. So we can in principle chart the evolution of dark matter structure with cosmic time. Measuring this growth of mass in our universe is a key goal, and it will ultimately lead to precision tests of theories of dark energy. To fully open this novel window of the 3-D universe of mass history, we need new telescopes and cameras very unlike what we have now.

What controls the development of structure in our universe? What do we know about the geometry and topology of our universe? What is the physical nature of dark matter and dark energy? These are fundamental questions that can be addressed through observation. Today, we have powerful new tools to test our theories of cosmology. The recently launched MAP satellite, as well as innovative ground-based radio interferometers and future satellites, will measure the geometry of the universe with precision. At optical wavelengths, arrays of charge-coupled devices are improving photon collection efficiency, and telescope apertures are growing. Together, they have increased photon collection by a factor of 10,000 over the past decade or so, capturing feeble light from the edge of the optically observable universe. Wavelength coverage has widened by an even larger factor. Wide-angle gravitational lens surveys, probing the cosmos in new ways, will generate millions of gigabytes of data and intriguing new opportunities for understanding the development of cosmic structure.

219

The wide and deep views of the planned Dark Matter Telescope will allow astronomers to chart the distribution of dark matter and detect the presence and influence of dark energy.

These advances in astronomical technology have equipped us to mine the distant galaxies for data—in industrial quantity. Our challenge is twofold: These galaxies are faint, and we need to capture images of billions of them. How can we do this? After a recent nationwide assessment, a community of professional astronomers recommended a dedicated facility combining large light-collecting capability and unprecedented field of view. The two-billion pixel camera of the planned "Dark Matter Telescope"(also known as the Large Synoptic Survey Telescope) will give astronomers a wide and deep view of the universe, allowing us to conduct full 3-D mass tomography to chart not only dark matter, but the presence and nature of dark energy as well. We will be able to map out the development of mass structure with cosmic time. Combining these results with other cosmic probes will lead to multiple tests of the foundations of our model for the universe.

What will our concept of the universe be when those answers are in? In a sense, the most interesting outcome will be the unexpected. For instance, a clash between different precision measurements might prove to be a hint of a grander structure, possibly in higher dimensions, and new mysteries.

DETECTING THE SIGNATURE OF THE BIG BANG

Robert W. Wilson

In science—just as in life—the people who stumble upon discoveries are not always the people who are looking to make them. And that helps explain how two scientists working for the phone company discovered the first hard evidence of the big bang.

In the 1920s Edwin Hubble found a correlation between distances to galaxies and the amount of redshift in their spectra. That redshift was evidence that the galaxies are flying away from one another—and it became the basis for several new theories of how the universe is put together. Theorists had used Albert Einstein's famous relativity equations to create various types of universes; some without mass and changing in time, and some with mass and changing in time. But after Hubble's

The first 300,000 years of the universe is visible at right. From the big bang (far right), the universe expanded and cooled to the glow of microwave radiation—the cosmic microwave background radiation—300,000 years later (far left). The big bang set in motion the evolution of the universe: galaxies, stars, planets, and people.

discoveries, any speculation about the nature of the universe had to include the fact that it was not static but expanding. The galaxies were flowing away from one another. There was no center to this expansion, and no known edge.

Naturally, Hubble's observations and the theory based upon them implied that, at one time in the distant past, the universe was infinitely dense and infinitely hot because expanding systems usually cool and grow less dense over time. This infinitely hot and dense time has come to be known as the "big bang," the explosive moment when the physical universe came into being. By adding in a few more assumptions, scientists can calculate not only how the temperature dropped in the early universe, but what kinds of elements formed in the process. Since the late 1930s, about ten years after Hubble's discoveries, theorists have been able to apply what was known about nuclear fusion to show not only what makes the sun and stars shine, but also what elements formed in the first moments after the big bang. George Gamow, Ralph Alpher, and Robert Herman did just that in the late 1940s. As a by-product of that effort they showed that the primordial fireball—or "cosmic egg," as some then called the big bang—could even be detected if a sufficiently sensitive radio antenna could be built. At that time, however, the attempt was not made.

Not everyone was comfortable with the idea that the universe began in an explosive event and has been changing ever since; there were many problems to overcome before this theory gained general acceptance. For instance, the age of the universe derived from the expansion rate Hubble had found turned out to be less than the calculated ages of the Earth and the stars, especially that of orbiting binary star systems and large star clusters. But as scientists measured more and more galaxies' distances and "redshifts" (which indicate just how fast they are moving), the numerical value of Hubble's expansion law changed, and the estimated age of the universe changed with it, generally upward. A big step toward the reconciliation of stellar and geological lifetimes came in 1952 when Walter Baade more than doubled the scale of the universe with a major recalibration of the distances to galaxies. Baade's research indicated that the universe was expanding much more slowly than had originally been thought—about half as fast. That meant it had taken longer for it to reach its presently observable size, so estimates of the age of the universe increased from Hubble's initial range of a mere 2 to 5 billion years to a more acceptable range of 10 to 20 billion years.

But reconciling ages was not the only concern astronomers harbored about the big bang theory. Some preferred that their universe be infinite in age, or at least of indeterminate age. Using Einsteinian relativity equations, in the late 1940s British mathematical theorists Hermann Bondi, Thomas Gold, and Fred Hoyle argued that the universe was homogeneous in space and time and had been like that forever—in other words, they believed the universe was in a "steady state." Matter, they argued, was somehow created evenly over time and space to keep the average density of the universe constant. This was not a new idea. Variants on this theme had been put forth by prominent physicists for almost 30 years by that time, the most provocative of which was that of Robert A. Millikan who, in the 1930s, proclaimed that "cosmic rays"—the enigmatic radiation from space first detected by Victor Hess in 1911—pervaded all space and were, in fact, the "birth cries of the elements."

In the 1950s and early 1960s, students of astronomy were usually taught the evolutionary "big bang" and static "steady state" scenarios side by side. The first edition (1964) of George Abell's popular textbook *Exploration of the Universe* gives them almost equal weight, suggesting numerous tests one might use to discriminate between the two theories. By the time of Abell's second edition (1969) the steady state theory was still discussed but its significance was placed in the past tense and the discussion of it appeared in an entirely new section on "Observational Evidence Against the Steady State." Abell presented a number of lines of evidence. First was Maarten Schmidt's analysis of the local rarity of a highly luminous class of galaxy called quasi-stellar radio sources or "quasars," which implied an evolutionary universe. Then Abell recounted what he called "one of the most exciting discoveries in recent decades," a discovery in which I had the good fortune to play a part.

I was one of the two Bell Labs scientists who, in 1964, found the remnants of the big bang. Even after we made this discovery, it took us quite a long time to realize that what we had found was evidence of the event that brought our universe into being.

In 1963, after completing my graduate training at Caltech, where I had surveyed the Milky Way with a large radio antenna to try to detect various parts of its structure, I accepted a job at Bell Labs to find ways to refine how we used radio telescopes for astronomy. In my thesis work with John Bolton, who had recently come to Caltech from Australia, I came to realize that there were limitations to the radio technique that kept me from detecting the components of the galaxy that people were then speculating about. Specifically, did our Milky Way have a huge spherical halo of radiating material surrounding it? Neither the sensitivity of radio telescopes that were then available, nor general knowledge about the radio universe were complete enough to allow anyone to discriminate among the many sources of radiation that were mixed together. During my graduate years I had become familiar with the work at Bell Labs on the new maser amplifiers, which were not only being built for Cold War surveillance radars, but which also had important spin-off uses for radio astronomy. I also knew that a small radio antenna at Bell Labs was very well suited to sorting out the many sources of radiation. So, upon learning that Bell Labs was looking for a few radio astronomers, I signed on and moved to Holmdel, New Jersey.

Since the late 1950s, Bell Labs, at its Holmdel site, had been actively seeking ways to build and maintain sensitive global communications systems using communication satellites. The idea of using satellites for communications was not new; Arthur C. Clarke had made his reputation by suggesting such a scheme in the 1940s. The Echo satellite was the first such Bell project; this gigantic Mylar balloon that NASA was going to place in orbit could, among other things, act as a reflector of radio signals. Knowing that balloons make for poor reflectors, the Bell scientists concentrated on making their receivers as sensitive as possible. In other words, they looked for ways to reduce "noise," the signals coming from things other than the object on which the device is focused.

Bell Labs was a great place to find the best equipment. In addition to its state-of-the-art "ruby traveling wave maser" low-noise receiver, the company had a new type of collecting system—the horn reflector antenna. It is now a familiar sight because it has been adopted as a standard for the microwave towers that have popped up all across the country. One of these horns, mounted on the ground, will pick up very little "noise" from the

THE CENTER IS EVERYWHERE

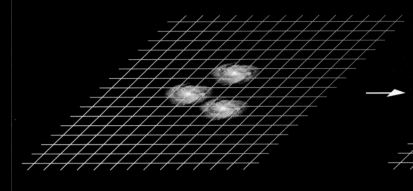

Ever since Edwin Hubble found that galaxies are all moving away from us, people have wondered if this does not mean that we are the center of the universe after all. Actually, what it means is that there is no preferred center. The universe is centerless, or "acentric." There is no preferred direction to the big bang, because we are inside of the big bang, and everything we can see in this universe is also part of the original big bang. Yes, you might say, but why are all the galaxies still moving away from us?

This is where an understanding of Hubble's work comes in handy. The Hubble law for the expansion of the universe says that galaxies are moving away from us at speeds proportional to their distances. If a galaxy is at a distance we'll call "D" and is found to be moving away at speed we'll call "V," a different galaxy at distance twice D, or 2D, will be moving at twice the velocity, or 2V. Furthermore, Hubble's law works whatever direction we look. The expanding universe is like the surface of a balloon that is being inflated. If we put marks on a balloon's surface to represent the galaxies and then inflate it, the marks will all move farther and farther away from one another as the volume of the balloon increases. So, if we are sitting on one of the galaxies and look left and right, we will see other galaxies moving away from us in proportion to their distance from us. It looks like we are at the center of this universe. This concept is illustrated below.

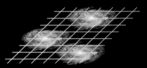

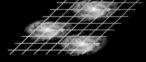

Now imagine the same universe, but with THEM as the observer. Remember, the observer is always at rest. When THEM looks at US they see us moving to the left at speed –V. When they look at the galaxy to our left,

US　　　THEM

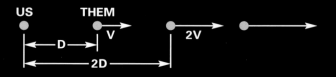

they see it moving left at –2V, relative to THEM.

If you continue the argument, you will see that THEM sees a Hubble universe cen-tered on THEM. In a universe that expands according to Hubble's law, it appears to observers that they are at the center.

—David Wilkinson

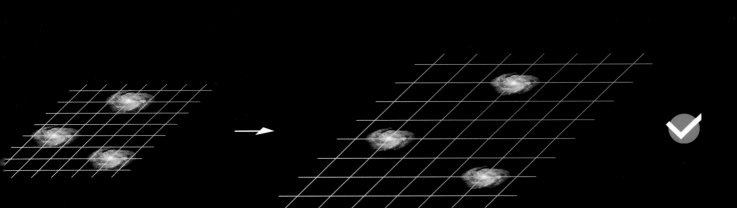

Misconceptions about the big bang abound. TOP: Galaxies are flying through space, scattered by the big bang (wrong). MIDDLE: The galaxies are expanding along with space (wrong). BOTTOM: Space itself is expanding, carrying the galaxies along with it (correct!). The big bang was not an explosion into space, but an explosion of space.

ground when it is pointed at the sky.

An excellent 20-foot-wide horn reflector that had been built by A.B. Crawford as part of the satellite communications program already existed on Crawford Hill at Holmdel. The receiver was located in the cab at the neck of the horn and was well shielded from the rest of the environment. Much smaller, but far more discriminating than the familiar, open, bowl-like structure of a parabolic antenna, it made for an excellent low-noise antenna. These two devices were a perfect combination with which to explore ways to improve satellite communications systems.

There was one other radio astronomer at Bell Labs when I arrived, Arno Penzias, who had come in 1961 as a member of the technical staff. He was already conducting research in radio communication and had worked on the Echo and Telstar communications satellites. Arno had come to America with his parents in 1940 from Munich, barely escaping extermination by the Nazis. He studied physics at the City College of New York, and received his Ph.D. in physics from Columbia under Charles Townes, the inventor of the maser and a familiar figure at Bell Labs. Arno had also served in the U.S. Army as a radar technician. At Bell Labs he was given the task of building a maser amplifier—his specialty—under Townes's direction, and was also encouraged to use it for radio astronomy. "The equipment-building went better than the observations," he once recalled.

But why would Bell Labs want to hire two astronomers?

Furthermore, why would two astronomers want to start off with such a small antenna? I think the answer for Bell Labs was two-fold: a desire on the part of the people who had built this very low-noise receiving system to see it put to good use for science, and an enthusiasm for improving satellite communications. One way to address both of these objectives was to hire some people to do science in their area of interest. As it turned out, Arno and I were those people. Both of us were enthusiastic about the 20-foot horn reflector because of its specific capabilities. It is a small enough antenna so that you can calibrate it and keep it calibrated, yet sensitive enough to measure a reasonable number of extraterrestrial sources. We realized that it was possible to make direct, or "absolute flux," measurements of radio sources. This is something that astronomers do not often do; they usually calibrate sources with respect to other sources and so obtain only relative intensities. Sources are also measured relative to their background, first pointing the antenna at the source of interest and taking a reading, then pointing it at some adjacent piece of empty sky, and finally taking the difference between the two readings. Now we could measure intensities in absolute terms, as we would in a physical laboratory. To understand the absolute radiation characteristics of any source, we had to understand our antenna better than astronomers typically did at the time. As I said in a conference in May 1983 at the National Radio Astronomy Observatory in Green Bank, West Virginia, "Radio astronomers don't often understand the background temperature [sources of noise] when they do the usual experiment of pointing at a source and pointing away from a source." That was how radio astronomers made observations, but I knew from my experience at Caltech that there was much room for improvement. And that is why I was attracted to the absolute calibration work which could be done at Bell Labs.

Arno and I teamed up to make the 20-foot horn useful for astronomical observations.

With its set of electronic amplifiers and receivers, especially the ruby traveling wave maser, it was potentially the most sensitive instrument in the world. Its well-shielded design also made it one of the most discriminating of antennae, able to sort out subtle differences in radio noise sources.

Arno was already well along in converting this equipment for astronomical measurement when I arrived at Bell Labs. He knew that the first step was to calibrate the receiver's temperature scale, and for that purpose he really overdid the job! He built what is known as a "cold load"— basically an extremely cold, and therefore low-noise, calibration source. It was a cylindrical nest of wave-guides, gas baffles, nitrogen pre-coolers, and gaseous and liquid helium containers and absorbers. All this was wrapped in a "dewar"—essentially a highly efficient, well-insulated thermos bottle. We poured some 25 liters of liquid helium in it and calculated the radiation temperature at the top to be approximately 5 kelvins. That's an extremely low temperature, close to absolute zero, well over 400 degrees below the freezing point of water on the Fahrenheit temperature scale.

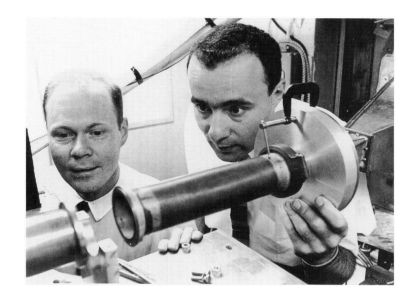

In 1964 Bell Labs scientists Robert Wilson (left) and Arno Penzias discovered, quite by accident, that the entire sky seemed to be glowing with microwave light.

Arno handled most of the cold load preparation while I set up our radiometer, the device that actually measures the temperature of things, like an extremely sensitive electronic thermometer. Radio telescopes actually measure the temperature of radio sources. The sky has a temperature, the moon has a temperature, and so does anything that has any heat at all, even if that heat is what we would describe as "cold." The radio spectrum is simply an extension of the spectrum of visible light that we are familiar with. The electromagnetic spectrum, as it is known to scientists, has high-energy gamma rays and x-rays at one end, the visible region as a narrow wedge in the middle, and the microwave and radio spectrum stretching out at the low energy end of the spectrum. Every physical body in the universe that has any temperature at all will radiate some energy across this spectrum. The coolest objects will be "brightest" in the low energy range—the radio and microwave. Radio astronomers and communications engineers call these thermal sources noise because they are incoherent and cause fluctuations (such as snow in a TV set) even if their average value is subtracted out. Our job was to account for all of the noise sources so that their average value could be subtracted from our measurements of the horn to give the temperature of the cosmos beyond the Earth.

There were indications all along that sources of radio radiation were not being accounted for. After the 20-foot horn was built and was being used with the Echo satellite, Ed Ohm, who was a very careful experimenter, added up all the components of the system and compared that figure to his measured total. He had predicted a total system temperature of 18.9 kelvins, but he found that he consistently measured 22.2 degrees, some 3.3 kelvins more than he had expected. However, that was within the measurement errors of his instrumentation, so he didn't consider the discrepancy to be significant.

I distinctly remember that, as Arno and I put the helium-cooled reference source on the 20-foot reflector and got everything working, we knew that we were going to be either happy or sad with the result. The antenna's measurement of the reference source's temperature would either be within what we considered an acceptable range, based on our calculations of the "noise" it would also register, or it wouldn't.

Our first observations were disappointing. We had hoped that the discrepancies could be explained by the known limits of our instrumentation. Our observing technique, based on one developed by Bob Dicke, was to compare the antenna's temperature to that of the the cold load.

Penzias and Wilson used the huge horn-shaped antenna (foreground) at Bell Labs in Holmdel, New Jersey, to discover the cosmic microwave background radiation—critical proof of the big bang theory.

When the antenna was pointed straight up, the radiation temperature was about 7.5 kelvins. The problem was that we had expected 2.3 kelvins from the sky, and possibly 1 kelvin from the absorption in the walls of the antenna. We saw something that was considerably more than that—the antenna's reading was hotter than the helium in the cold load and it should have been colder. Clearly we had a problem, and it lay either in the antenna or beyond.

We knew that unless we could get to the bottom of this mystery we would not be able to do an experiment I had wanted to do since my days at Caltech: determine whether the galaxy was surrounded by a halo of radiation. Arno's immediate reaction was, "Well, I built a pretty good cold load, since imperfections in it would have the opposite effect," so we looked for other sources of error.

Maybe it was the Earth's atmosphere. At the time of our experiments, many radio astronomers thought that the radiation of the Earth's atmosphere, visible in the centimeter range of the spectrum, was about twice what we were finding. That would have gone a long way toward explaining our problem. We could, however, unequivocally rule this out from measurements of the equivalent temperature of the horn at several look angles lower than straight up. Conventional antennas were not so well suited to this measurement.

Maybe it was New York City. Crawford Hill, the site of the radio horn, overlooks the city of New York, which produced a lot of interference. We turned our antenna down and scanned the horizon, scanned below the horizon, above the horizon, and we found a little bit of man-made radiation, but nothing that would explain what we were seeing.

Maybe it was the Milky Way. We considered this possibility, but soon realized that the characteristics of the radiation we were detecting did not fit at all with what the Milky Way would look like if one extrapolated its known energy distribution into the range we were observing. Radiation from the actual plane of the Milky Way fit very well with what we expected it to be.

Maybe it was many discrete sources. No known population of sources could explain it. The strongest discrete radio source we could see was Cassiopeia A, believed to be a remnant of a supernova, and it could not account for the discrepancy as it was only twice as hot in our antenna at its peak as our excess.

Maybe it was the pigeons. We knew that pigeons were roosting in the horn, so we thought maybe their droppings radiated enough to create the problem. We set Have-A-Heart traps and evicted the pigeons, then swept out the antenna, cleaned it thoroughly, and sealed all the joints with aluminum tape. No change.

Maybe it was the Van Allen radiation belts. In 1962, the United States conducted a high-altitude nuclear explosion that had filled two already highly charged band-like regions in the Earth's atmosphere (the so-called Van Allen radiation belts) with additional high-energy charged particles. The belts were still calming down during the course of our observations, and during that time we continued to notice no change. So it was not the belts.

We lived with this problem for about a year while understanding our system and completing an accurate measurement of the flux of Cassiopeia A. If we couldn't understand our system at our first frequency of 4 GHz where a galactic halo should be very weak, we had no hope of measuring it at 1.4 GHz where we hoped to detect it.

We were really scratching our heads about what to do until one day when Arno was talking on the phone with Bernie Burke about other matters, he happened to mention our results and our inability to determine what was going on.

Burke told Arno that a group at Princeton was working on something that might explain what we were seeing. Robert Dicke, the leader of Princeton's physics group, who had developed a terrific switching device for microwave radiometry and was now interested in cosmology, had made a fascinating prediction. He suggested that if we live in an oscillating universe—one that expands and then contracts back to another super-hot big bang—it will, at the time of each big bang, cleanse itself of all heavy elements and begin the expansion over again, repeating this cycle throughout eternity. One outcome of this cyclic behavior is that in each cycle the universe will eventually relax into a state of thermal equilibrium. The expansion rate of the universe, and its subsequent cooling, causes the thermal spectrum to be Doppler shifted from the highest gamma ray range all the way to the other end of the spectrum—into the microwave and radio range. Although this "reddening" will stretch out the spectrum, it will still look like a thermal spectrum.

229

Dicke realized that this radiation might be visible in the microwave range. His experience with radiometers had taught him that the thermal radiation left over from the big bang was something that he could look for, and the understanding that this was possible made an enormous contribution to astronomy. Prediction or not, this notion gave Dicke's graduate students a good reason to build a sensitive radiometer, which they were well along with doing in late 1964, just when we were struggling with our own radiation-detection equipment. There had been other predictions of the thermal signature of the big bang, and even some obscure astronomical measurements, but these were unknown to us, and to Dicke's group.

When Burke told Arno what the Princeton group was up to, Arno called Dicke. This led to visits back and forth, and eventually to our realization of what it was we had discovered—the telltale radiation from the big bang. Still, we made one final check. We took a signal generator, attached it to a small horn, and took it around the top of Crawford Hill to artificially increase the temperature of the ground and measure the characteristics of our antenna sensitivity in the backfield. We did not find anything unexpected. Only then did we send a notice to a journal to formally tell the world what we had observed.

Arno and I were very happy to have some sort of answer to our dilemma, but I don't think either of us took the cosmology very seriously. Although I had come from Caltech, where the prevailing view was toward the big bang, I leaned toward the "steady state," and I was quite happy with that theory. As a result, I thought our measurement might survive any possible theory explaining it, and keeping our findings separate from any claims about the big bang seemed like a good idea, especially since the steady state people might come up with an equally good explanation.

Was there a sense of discovery? Yes, but it was very, very gradual. The whole thing was a puzzle for a long time and the cosmology was not convincing to me at first. In fact, I think it wasn't until Walter Sullivan wrote about our findings on the front page of the *New York Times* that I took the matter seriously. Only then did I realize that the rest of the world was taking it all very seriously, including the cosmology.

Our detection gave the world the first hard evidence that there had been a big bang and allowed some measurement of properties of the early universe. It gave us enormous recognition, of course, and changed the way we thought about astronomy and cosmology. However, we still worked at Bell Labs, which has a terrific tradition of scientific research, but after all is a subsidiary of a phone company. One day our boss's boss came in and, in effect, said, "You guys have been doing radio astronomy full time; the effort is supposed to be about half-time. Let's get on with something for the telephone company." We complied, but we had

achieved considerable equity by then. The astronomical community immediately took to our findings, and there were lots of other jobs for us, and for lots of other people, and scientific papers were coming from all directions. The world of astronomy now had a new and important physical phenomenon to explore, a phenomenon that held yet other keys to unlocking our cosmological origins.

Penzias and Wilson ruled out every imaginable source for the faint static affecting their sensitive radio antenna. They even trapped and removed pigeons roosting in the antenna horn, on the chance that heat from pigeon droppings was the cause. The static turned out to be something else: the fading flash from the big bang.

DETECTING THE SIGNATURE OF THE BIG BANG

THE BIG BANG AND ITS FIREBALL

David Wilkinson

When I came to Princeton in 1963 to work with Robert Dicke, a physics professor at the university and one of the world's leading experimentalists, one could learn what there was to know about cosmology in a week or two. There was still a vigorous debate about whether our expanding universe evolved from a very hot, condensed state (the big bang model), or whether it has always looked as it does now and always will (the steady state model). How was it possible that such an important issue could still be debated 34 years after American astronomer Edwin Hubble discovered that the universe is expanding?

I believe the answer lies with the relatively crude technology—mainly detectors, electronics, and computing—that was available to astronomers in those days. By the mid-1960s optical

Ripples in the cosmic microwave background radiation reveal the beginnings of large-scale structure in the newborn universe. The colors indicate tiny temperature differences—about one ten-thousandth of a kelvin—in the radiation, which is a frigid 3 kelvins above absolute zero.

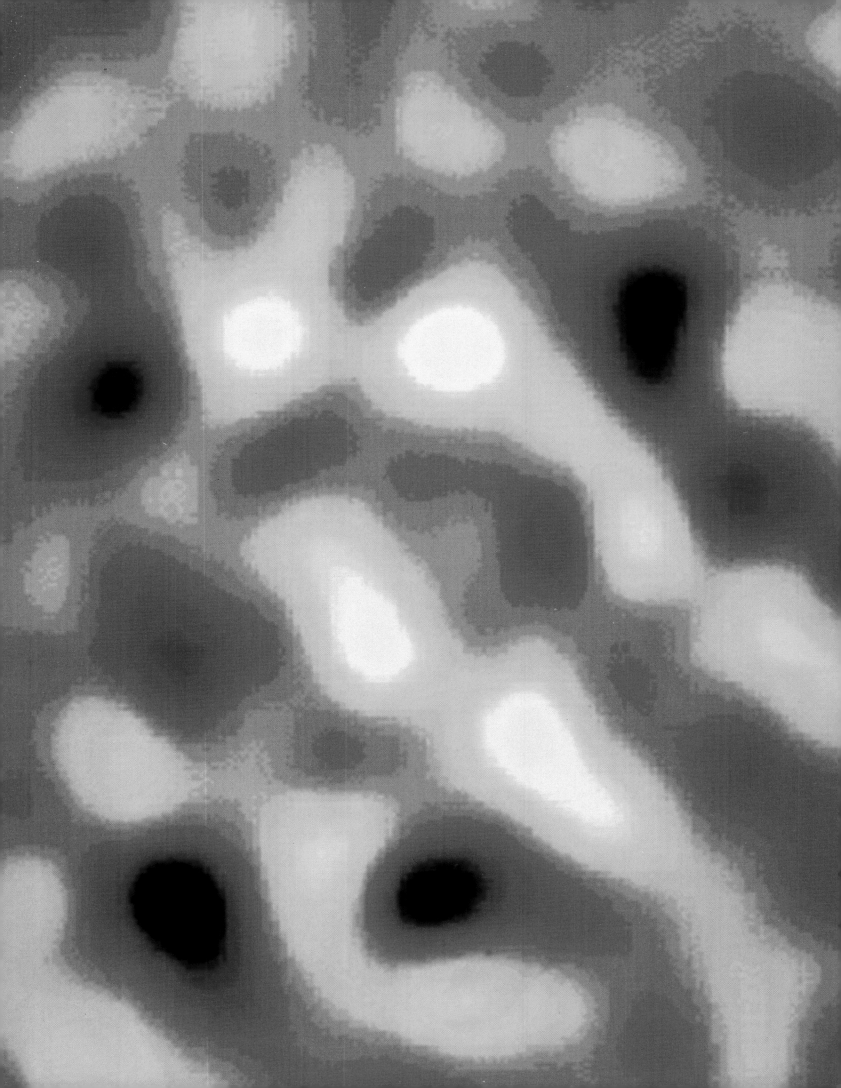

astronomers were using the new, post–World War II technology, including radio astronomy and bigger optical telescopes—especially the 200-inch telescope on Mount Palomar in southern California—to make careful measurements of the distances and recession speeds of distant galaxies. The goal was to discover the fate of the universe. Would it expand forever, or were the gravitational forces between galaxies slowing the expansion sufficiently to lead to eventual collapse? We know now that this program gave misleading results because the intrinsic brightness of galaxies changes as one looks deeper into the universe, leading to erroneous measurements of galactic distances. The main problem with the pre-1965 cosmology was that the field was starved for measurements. This changed over the next three and a half decades as more-sensitive detectors, bigger telescopes, faster computing, and astronomical satellites were brought to bear on cosmological problems. In 1965 the discovery of an important, but previously unknown, component of our universe—the cosmic microwave background radiation, or CMBR, marked a milestone in modern cosmology. Though initially controversial, the interpretation of this radiation as the fireball from the big bang itself has withstood the tests of time and intense scientific study. My introduction to the CMBR idea came in 1964 when my mentor, Robert Dicke, outlined an idea he had about a big bang version of cosmology. More precisely, he was thinking about an oscillating universe with successive cycles of expansion and contraction. His theory went like this: We see that the elements heavier than helium are produced in stars and supernovae explosions. If this continued for several cycles of expansion and contraction, the matter in the universe would become mostly heavy elements. But observations show that most of the matter in the universe is hydrogen, the lightest element. Dicke reasoned that the heavy elements must be *evaporated* in a "big crunch" when the universe collapses to a small size. Temperatures greater than ten billion kelvins are needed to disassemble the nuclei of heavy elements, so the universe must have reached a temperature of ten billion kelvins or more in the crunch.

As the universe expanded and cooled after the big bang, wavelengths of the fireball radiation would be stretched out until they were primarily in the microwave band, as they are today, thus the name "cosmic microwave background radiation." Dicke suggested that we build an instrument to look for evidence of microwave radiation from the fireball. In 1946 he had invented the instrument that was needed for the job—the Dicke radiometer. We made two special modifications to the standard instrument. First, we built what is called a "cold load"—a very stable source of liquid helium at extremely low temperature that could act as a calibrating, or reference temperature source for the radiometer. Next, we needed a specially designed "horn antenna" to collect radiation from the sky and reject radiation from the surrounding ground. Both of these components were hallmarks of any experiment to detect the "cold" CMBR in the midst of "hot" thermal noise. Dicke's idea was intriguing, but the underlying argument was less than compelling; it contained many assumptions and loopholes. Experimental physicists usually need stronger motivations to invest in an experiment that will take several years to complete, especially one that might end up providing evidence of nothing but the absence of a detectable signal, and consequently be uninteresting to others. Fortunately for me, I had nothing else to do at the time, so I signed on, along with my young colleagues Peter Roll and Jim Peebles. Roll and I started to build the radiometer, and Peebles started doing calculations to look for firmer ground for the experiment to stand on. One of his projects was to calculate whether some heavy elements would be made as the fireball temperature cooled to below ten billion kelvins.

Each Tuesday the four of us would gather in Dicke's office to eat lunch and discuss progress on the "fireball project." In the spring of 1965 one of our discussions was interrupted by a phone call that would change everything. Dicke was always getting phone calls, but this particular one got our attention when we heard him talk to the caller about a "cold load" and a "horn antenna," just the devices that we were developing, the two special components of a radiometer designed to look for the CMBR. When Dicke hung up the phone, he turned to us and

In the early 1960s, Robert Dicke used a radio detector at Princeton University to search for background radiation in the sky. The author, formerly one of Dicke's graduate students, stands at center-right.

said, "Well, boys, we've been scooped." The call came from Arno Penzias and Bob Wilson, radio astronomers at Bell Labs, who had been measuring, and trying to track down, excess noise in their radio telescope. Their work had been so careful and thoroughly checked that, after a few minutes of conversation, Dicke was sure that they had discovered the CMBR. On a subsequent visit to Bell Labs at Crawford Hill, only 30 miles from Princeton, we looked at the data, and Dicke described his idea for the CMBR. Two papers, which changed cosmology research, followed shortly.

I am often asked how it felt to be scooped on an important discovery. I was disappointed that our experiment wouldn't be the first to see the cosmic fireball, but now I could imagine a whole new field opening up, and I was there at the beginning. That was exciting. When the Nobel Prize for Penzias and Wilson was announced, I remember wishing that Dicke had shared it. He had had a wonderful idea. However, there was more to the story.

As so often happens in science, Dicke's idea was not new. George Gamow, Ralph Alpher, and Robert Herman had predicted the CMBR 17 years earlier. Gamow had the idea that the elements heavier than hydrogen might be *produced* in the hot, dense stages of the big bang, when spontaneous nuclear reactions could take place. (Dicke's premise was that the heavy

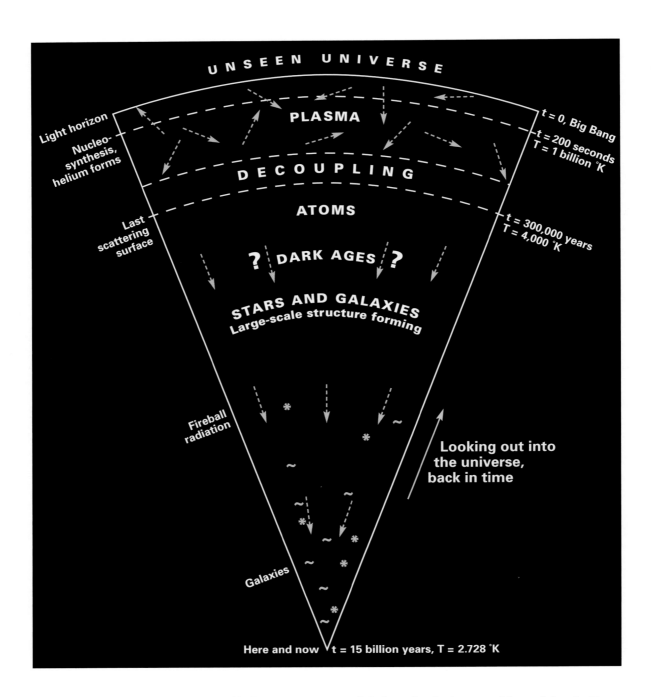

UNSEEN UNIVERSE

PLASMA

Light horizon

Nucleo-
synthesis,
helium forms

DECOUPLING

ATOMS

Last
scattering
surface

? DARK AGES ?

STARS AND GALAXIES
Large-scale structure forming

t = 0, Big Bang

t = 200 seconds
T = 1 billion °K

t = 300,000 years
T = 4,000 °K

Fireball
radiation

Looking out into
the universe,
back in time

Galaxies

Here and now t = 15 billion years, T = 2.728 °K

elements would have *evaporated*.) Gamow's group did the calculations and found that helium and a few other light elements could be produced a few minutes after the bang, but not the heavier elements. The main goal of the project was not successful. However, a by-product of their calculations was the presence and persistence of thermal radiation, the big bang fire-

Because light travels at a finite speed, objects appear to us as they did when the
light we see was emitted. The farther away objects are, the younger they appear.
By looking deep into the universe, we can literally look back in time—nearly all
the way back to the big bang itself.

ball. They calculated that the radiation should still be around today and have a temperature of 5 kelvins above absolute zero. Why, for 17 years, no one went out to look for this remnant of the big bang remains a mystery. The Gamow group had predicted an important new component of the universe, which needed to be checked.

Our thinking in 1965 went something like this: First, the CMBR offers strong evidence that we live in a big bang universe. The fireball is a natural consequence of big bang cosmology, but the steady state theory does not account for thermal radiation filling the universe. If measurements of CMBR agreed with detailed predictions of the big bang model, the steady state theory was in serious trouble. Second, cosmologists now had a new component of the universe to study, one that carried information from the very early universe, before stars and galaxies formed. The stakes were high, and astronomers and physicists were eager to test the fireball hypothesis.

The big bang theory predicts two necessary properties of the CMBR. Because the radiation was born in a hot fireball, it should have a particular spectrum. That is, the intensities at different wavelengths should have a certain shape, called the blackbody spectrum, or thermal spectrum. Measurements of the intensity of the CMBR at different wavelengths should fit onto a blackbody curve. Penzias and Wilson and our team from Princeton used radiometers to measure the CMBR temperature at two different wavelengths, which gave a crude measurement of the CMBR spectrum. The result, that the CMBR had a temperature of about 3 kelvins above absolute zero, fit the shape of the blackbody curve. A second issue was the distribution of temperature across the sky: The radiation should have the same temperature when measured in all directions in the sky if the big bang was initially completely uniform. Both radiometers were used to scan the sky. They showed no change in the radiation temperature to an accuracy of a few percent. The big bang explanation of the CMBR began to look pretty good. Further support came from an unexpected quarter. Several astronomers remembered an optical observation made in the 1930s that had indicated that interstellar molecules of cyanogen (a molecule common in cold astronomical sources) behave as if bathed in thermal radiation of about 3 kelvins. This is exactly what is expected if the CMBR has a blackbody spectrum and is everywhere in space. At this point, the CMBR measurements were attracting a lot of attention.

We knew that several other research groups were now actively searching for better techniques and technologies to improve measurements of the CMBR. At Princeton we continued to develop radiometers for use on the ground that would trace out the CMBR spectrum as well as determine how uniform the radiation was over the whole sky. We searched for the latest technology to reduce the electrical and thermal noise in our detectors, but, as our detectors got better (less noisy), we found that interference and noise from local sources on and around the Princeton campus became major problems. The student radio station, laundry truck radios, and police calls frequently overwhelmed our cosmic signal. We used every trick in the book to combat the effects of fluctuations in atmospheric radiation and the (relatively intense) radiation from the ground itself, but in the end they were insufficient. So we turned to balloons, taking advantage of the unique services of the National Scientific Balloon Facility in Palestine, Texas. We built a series of balloon packages and took them to Palestine or

Hobbs, New Mexico, for launch. A typical flight would launch at sunset, reach float altitude an hour and a half later, then fly all night and parachute to the ground at sunrise. A recovery airplane and crew would be waiting, having chased the balloon into Mississippi or Alabama. Wads of 20-dollar bills were used to reimburse farmers for crops, cows, and sometimes building damage. Our gondolas included special screens to block the radiation from the ground below. At 90,000 feet, the thin atmosphere above was not a problem.

One gauge of the sensitivity of a radiometer is the time needed by the system to detect weak signals. Experiments to detect irregularities in the CMBR need to reach a signal strength of 0.001 kelvins and smaller. The original Princeton radiometer was terrible; it took ten days to measure signals of 0.001 kelvins. By 1970 we had a radiometer that took only three hours; and by 1996 my colleague Lyman Page and his group were flying a cryogenic radiometer that reached a sensitivity of 0.001 kelvins in one second. These instruments used radio techniques. Cryogenic bolometers are another detector that now achieve an accuracy of 0.001 kelvins in a fraction of a second. We were always searching for more stable and sensitive radiometers, well beyond anything that had been achieved before. Detector designers took up the challenge, developing ever better sensitivity, often using new scientific and technical principles. Thus, an old story was played out on the small CMBR stage: The special needs of science motivate new technology and new technology enables new science.

In the 25 years following the discovery of the CMBR many measurements of the spectrum and temperature uniformity of the CMBR were made from the ground and from balloon platforms. The measured CMBR spectrum fit the shape of a blackbody radiation curve with progressively better accuracy, and the CMBR temperature was found to be the same in all directions. By 1990, most cosmologists had adopted the big bang theory as their working model. Of course, during that same period, astronomers were making observations of the universe at other wavelengths, mainly in visible light. Their results were also indicating that we live in an evolving universe, as implied by the big bang model.

To understand the big bang model, imagine a slice of our universe. We are at the point, or vertex, of the slice, looking out into the distant universe. But the distance that we can see is limited by the finite speed of light (300,000 kilometers per second) and the age of the universe (about 15 billion years). In principle, we can see no farther than the distance that light can travel in 15 billion years, or 15 billion light years. For reference, the Earth is eight light minutes from the sun; we see light from the sun eight minutes after it was emitted. According to the big bang model, a shell of radiation from the big bang, the CMBR, encircles the universe at a distance of 15 billion light years in every direction.

One of the most provocative questions we can ask is, What is outside of that shell in the region called the "unseen universe"? Presumably the same stuff as just inside, but we cannot know because there has not been enough time in the life of the universe for light (information) to get to us from outside. Our universe can be thought of as a sphere embedded in a much larger, perhaps infinite, unseen universe.

Another consequence of finite light speed is that, as we look deeper into the universe, we look farther back in time. We see cosmic light sources as they were when the light was emitted, not as they are when we receive the light. In this sense, telescopes are time machines. By studying distant galaxies, we can learn how galaxies first formed and how they evolved. Because galaxies contain billions of light-emitting stars, they can be studied with large optical telescopes. The larger the telescope, the deeper we can look. Our telescopes are not yet large enough to see the objects in the region where stars, galaxies, and large-scale structure are forming, but we are getting close.

When we observe the CMBR, we are seeing the universe as it was when the radiation last scattered from matter, the "last scattering surface." This happened at a time when the matter was changing from a sea of electrically charged particles (plasma) to a sea of neutral atoms. In the hot big bang model the matter becomes neutral when the radiation tempera-

Launched in 1989, the Cosmic Background Explorer (COBE) carried three Differential Microwave Radiometers (DMRs) similar to this radiometer, which had been flown earlier on a U-2 aircraft by scientists from the University of California, Berkeley. Each DMR measured tiny variations in the brightness of the background radiation.

Packed within this cylinder are prototypes of two other COBE instruments, FIRAS
and DIRBE. To measure the feeble heat from the background radiation, both had to
be kept colder than the heat of the radiation itself. The cylinder was mounted in a
large cryostat, which supercooled it to less than 3 kelvins above absolute zero.

ture drops below 4,000 kelvins, about 300,000 years after the bang. In this important cosmic epoch, called "decoupling," the radiation and the matter stop interacting. The CMBR carries information to us from the time when the universe was about one-thousandth of its current size and only a small fraction of its current age. It is like being able to see what a 60-year-old person looked like ten hours after being born. This was long before stars and galaxies formed.

At decoupling, the universe was much smoother than it is now because gravity was unable to clump the matter in its plasma state, so the CMBR is expected to be very uniform across the sky. However, theorists soon calculated that the CMBR cannot be perfectly smooth. There must be small fluctuations in the matter density to act as seeds for forming galaxies and large-scale structure later on. These matter density fluctuations are accompanied by fireball temperature fluctuations, which appear now as very weak irregularities in the CMBR temperature. An accurate map of the CMBR should show tiny irregularities in the temperature from place to place across the sky. Theorists predicted that the temperature would vary by about 0.00003 kelvins, one part in 100,000 of the 3-

241

John Mather's graphical presentation of the COBE FIRAS measurement data, which proved the big bang theory, sent a surge of excitement through the astronomical world.

kelvin CMBR temperature. Measuring these tiny variations presented a significant challenge to the experimenters, especially with instruments surrounded by 300-kelvin radiation from the Earth, which can easily contaminate the CMBR signal.

As mentioned earlier, Gamow's group, and later Peebles, predicted that, in a big bang, nuclear reactions form helium and traces of other light elements. These transformations happen about 200 seconds after the bang, when the radiation temperature is about a billion kelvins. Refined calculations have made accurate predictions of the amount of each element produced. These predictions play an important role in cosmology, because astronomers can measure the amount of these elements present in the universe today. The measurements agree with the predictions, once again lending strong support to the big bang model. We know very little about the epoch between decoupling and galaxy formation, conjectured to be between one and five billion years after the big bang. The Hubble Space Telescope and large ground-based telescopes are beginning to probe the cosmic epoch called the "dark ages," when stars, galaxies and large-scale structure were just starting to form. Observations by deep sky surveys, such as the Sloan Digital Sky Survey and the Australian 2DF surveys are leading the way.

It may be tempting to think that we are at the center of the universe, as the 15-billion-light-year "shell" model of the CMBR suggests. Of course, this is not so. In fact, observers at different places in the universe draw the same picture, putting themselves at the vertex of the slice. And because the optical horizons of different observers do not coincide, an observer one billion light years from us could see past our light horizon in that direction. Unless the unseen universe in that direction is the same as what is just inside our light horizon, that observer will see a very uneven, or "anisotropic," universe. When the Hubble expansion of the universe is included in the picture, it looks even more as if we are at the center, but this is not so. The center is everywhere

What is the evidence that persuades cosmologists that the big bang model describes our universe? Early measurements of the CMBR spectrum agreed with the predictions of the model, and the measured abundances of lightweight elements matched the results of physicists' calculations. However, cosmologists wanted stronger evidence. Early in the 1970s a few CMBR experimenters began to design a satellite that could avoid many of the problems we were having with ground- and balloon-based CMBR experiments. The result was the Cosmic Background Explorer (COBE) satellite, started by NASA in 1975 and launched in 1989. The COBE carried instruments to measure the CMBR spectrum and to search for irregularities of the radiation in the CMBR, two stringent tests of the big bang model.

Within weeks after the COBE was launched, it had produced a spectacular measurement of the CMBR. The COBE's measurements of the CMBR fit the predicted blackbody curve exceedingly well. Seeing this curve for the first time was a great thrill for me. My students and I had been making these measurements, one point at a time (and one graduate student at a time), for 25 years. Suddenly, here was the answer, crystal clear. We live in a big bang universe. John Mather of NASA's Goddard Space Flight Center was the person most responsible for the success of this measurement. When he showed the figure to an overflow session of the American Astronomical Society, the audience spontaneously rose and applauded, a rare event at scientific meetings.

The COBE's experiment to determine slight irregularities in the CMBR temperature took much longer to produce results because of the much weaker signal that had to be measured. More time was needed to diminish the effects of instrument noise. Finally, in October 1991, science team member Ned Wright told the team that he had found a weak, but significant, CMBR irregularity in the data. Many of us were skeptical because spurious effects had not been accounted for, and the official analysis had not yet detected the effect. No one on the team wanted to announce such an important result and have to retract it later, but the level of excitement rose and the analysis accelerated. Finally, after much internal discussion and debate, the team was ready to announce a detection of CMBR irregularity in April 1992. George Smoot, the principal investigator for the experiment, showed the results at a meeting of the American Physical Society. The magnitude of the irregularity was close to that predicted, and CMBR theorists were ecstatic. Quotes like "the holy grail of cosmology," "the face of God," and the "greatest discovery of the century" hit the tabloids, and the COBE enjoyed a moment of media fame. However, the real news was that the COBE's anisotropy measurements had verified an important prediction of the big bang model.

Meanwhile, optical observations were going ever deeper into the visible universe, and

242

The COBE satellite measured the intensity of the cosmic microwave background radiation—the big bang fireball—over a range of infrared and microwave wavelengths, indicating an extremely cold radiation. The resulting graphed curve (this page, center) looked like a blackbody spectrum, which relates the variation of the intensity of radiation (vertical axis) to wavelength (horizontal axis). Any object that absorbs all incident radiation, regardless of the wavelength, is called a blackbody. It also emits radiation wavelengths according to a formula that graphically takes a universal shape, regardless of its composition. A heated iron bar has the properties of a blackbody because all the energy it radiates is thermal energy. The solid line in the figure shows the universal shape of a blackbody spectrum. The position of the peak and the height of the curve depend on the temperature of the emitter, but the shape is the same for all blackbodies. The cosmic background radiation spectrum peaks at a wavelength of about 2 millimeters. Your body radiates heat with a spectrum that peaks at about 0.02 millimeters—a shorter wavelength, hence a higher frequency and a higher energy density.

The blackbody spectrum figured in two important 20th-century discoveries: quantum mechanics and big bang cosmology. In 1900 Max Planck developed the quantum to explain the shape of blackbody spectra. Classical mechanics explained the shape of the curve at long and short wavelengths, but failed to explain its shape around the peak of the spectrum. Planck found that by assuming that radiation energy comes in packets (quanta) he could calculate the measured shape of the blackbody spectrum. Quantum mechanics was thus introduced to explain the blackbody radiation spectrum. Fifty years later, George Gamow found that blackbody radiation was a consequence of the big bang model, and that the radiation would still be filling the universe, though it would be much colder now. Measurements of the spectrum of the CMBR, especially those from the COBE satellite, fit the spectrum of a blackbody at a temperature of 2.728 kelvins above absolute zero.

Verifying the blackbody character of the remnants of the big bang helped determine what the big bang was. Bodies that radiate energy with purely thermal properties look exactly like the radiation properties of the big bang. Comparison between the observations of Penzias and Wilson and the Robert Dicke group hinted at this, but the difference in the wavelength ranges was small. Other physicists studied parts of the radiation distribution at different wavelengths. Balloons, aircraft, and eventually the COBE filled in the picture. They confirmed that the big bang was a thermal process—the radiation came from an extremely hot body, not from some other source of radiation.

—David Wilkinson

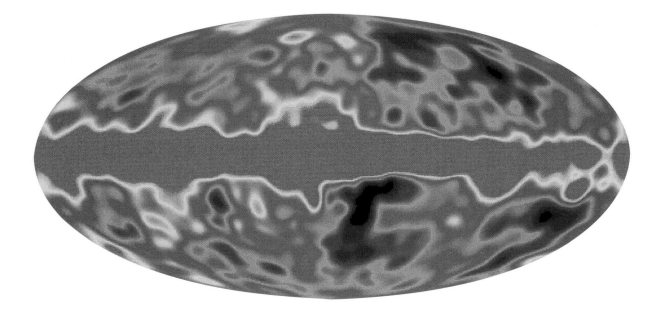

the results were supporting the big bang model. Evolutionary processes in galaxies were observed, and refined measurements of light element quantities still agreed with the big bang model. None of the new observations challenged the model, even though new phenomena were being found at an impressive rate. Also, during the lifetime of the COBE project, other satellites were opening new windows on our universe at infrared, x-ray, and gamma-ray wavelengths. New sources of radiation were discovered, some from deep in the universe. This wealth of new information supported, or was compatible with, the big bang model.

Once that the COBE team had shown fluctuations to exist, experimenters and theorists began to plan how to use this new tool to learn even more about our universe. Theorists had studied more carefully the epoch of the decoupling of radiation and matter and found that something unexpected might be happening at angles smaller than the COBE could see—spots 7 degrees in diameter, some 14 times the apparent angular size of the moon. If we could build radiometers to see spots two times the size of the moon and smaller, then we could test a detailed prediction of the big bang model—that fluctuations in spots of about 1 degree would be three times greater than those at 7 degrees.

Furthermore, a complex pattern of irregularities was predicted for angles of 1 degree and smaller. If the CMBR temperature variations could be measured at smaller angles, detailed physics of the decoupling epoch could be checked, and interesting properties of the big bang model could be measured. For experimenters, the need for smaller beams on the sky meant using larger antennas for collecting the CMBR, and detectors with greater sensitivity had to be developed. Initially, the search for peaks in the CMBR fluctuations would have to be done from the ground and from balloons. The old COBE results anchor the theoretical predictions at large angles. In 2001 the TOCO experiment, carried out on a 17,000-foot

plateau in Chile, was the first to show that there is a maximum intensity in the CMBR fluctuations around 1 degree. The BOOMERANG experiment enjoyed a spectacular week-long balloon flight around Antarctica. MAXIMA made a more conventional balloon flight from Palestine, Texas. DASI is a new kind of CMBR experiment, called an interferometer. To minimize problems from fluctuating atmospheric radiation, the experiment was done at the South Pole. These last three experiments were capable of measuring smaller spots on the sky than could TOCO. Perhaps they are beginning to see evidence for secondary peaks in the CMBR irregularities at angles smaller than 1 degree. But there is more to learn from CMBR fluctuations at smaller angles.

The correct shape of the theory-predicted CMBR pattern depends on important properties of our particular universe. How much of the total matter in the universe is ordinary stuff? How much is dark matter, which we know is present but have not yet identified? Will the universe expand forever or will it eventually collapse? These and other important cosmological questions can be answered by measurements of the precise shape of the temperature variations of the CMBR at small angles. Plans have been made to have two satellites do exactly that. NASA's MAP satellite was launched on June 30, 2001, and the European Space Agency plans to launch the PLANCK satellite in 2007. Both will be placed at Lagrange point number two (L2), a location about one million miles from Earth. At L2 the combined gravitational forces of the sun and Earth will cause MAP to orbit the sun once a year, thus keeping it always

Different colors in the COBE image (above left) indicate temperature variations in the background radiation. The next generation of satellites, MAP and PLANCK, will map the radiation at a much greater resolution than COBE, producing images similar to the one above, which will reveal in much finer detail the temperature variations present in the early cosmos.

in line with the Earth and sun. MAP is oriented to look away from the Earth and sun, and protective barriers shield its 20 radiometers from heat. Out of the sun, the radiometers radiate their heat to space and cool to about the same temperature as liquid air on Earth. Both the MAP and PLANCK satellites will make maps of the whole sky with much finer resolution than was possible with the COBE. After scanning the sky for two years, MAP will trace the patterns of the CMBR irregularities to find out which version of the big bang model best describes our universe.

Since 1965, cosmologists have had the good luck to see their research field develop from a scientific backwater into one of the most exciting areas of science. New technologies and access to space enabled measurements of the distant universe at all wavelengths of interest to cosmologists. New data from observatories, spacecraft, balloons, and aircraft has led to progressively more realistic and detailed cosmological models. I am happy to say that my field, the cosmic microwave background radiation, played a central role in this revolution in cosmology.

As an old-timer in CMBR research, I can appreciate how far we have come in 35 years. Progress in understanding our universe has been simply spectacular. It would be foolish to speculate on where cosmology research will take us in the next decades, but we can congratulate ourselves for having learned so much about the universe already, and from such a great distance. At 15 billion years of age, our universe is enormous in space and time compared to scales familiar to us. While far from being a simple place, it does seem to

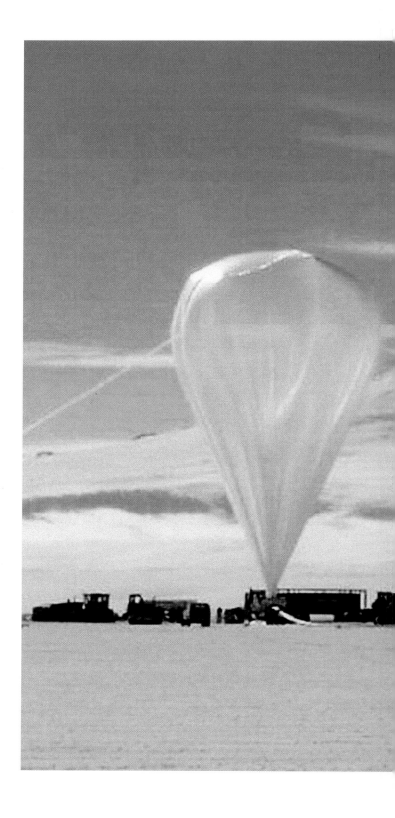

follow a set of rules, and we have been able to use our intelligence to learn those rules by thinking and experimenting in a very small region of cosmic space and time. The physics that we have discovered locally have been impressively effective in explaining complicated phenomena taking place in this vast universe. How fortunate we are that this is the case.

So named because prevailing high-altitude Antarctic winds carried it in a circle back to its launch site, balloon-borne BOOMERANG mapped subtle variations in the cosmic background radiation across a small area of the sky. Its cluster of detectors measured temperature differences as minute as 100-millionth of 1 kelvin.

THE BIG BANG AND ITS FIREBALL

COME! EXPLORE
THE UNIVERSE

David DeVorkin

This is an invitation to explore the universe at the National Air and Space Museum, which is free for the asking and open 364 days of the year. This is also an invitation to explore the universe at every opportunity. We at the Smithsonian hope that these essays have given you a stimulus to explore, and we hope that our efforts will help more people discover something new about the universe and about cultures that have pondered the universe in different ways. One lesson should be clear: you do not have to be an astronomer to explore the universe (but it helps). Physicists, art historians, ethnographers, and curators all touch the cosmos by looking at the universe in different ways.

In my experience, the sensory delights of astronomy are the attraction. Some of my earliest memories come from being driven by my father up into the Hollywood hills to look at the sky, catch a meteor trail, or, better yet, visit the Griffith Observatory and Planetarium, where you could actually look through a huge telescope. This telescope was magnificent; almost 20 feet long, in a heavy copper dome, it was both massive and mysterious. You usually had to stand in line outside the dome. Your first glimpse of the telescope, in very dimmed reddened light, was through a vestibule window. Then you climbed a few steps into the observing chamber where about 20 people could stand, waiting their turn. There was a small office tended by the telescope operator and every so often he would come out, check the instrument, which had things whirling around on it, and maybe even answer a question or two. If you were really lucky, it would be the time to move the dome, which the operator did by throwing a switch. "Oooh," everyone would gasp. Finally, it was my turn to climb a ladder to the eyepiece, where I could stretch to look into a large glass-ended brass tube straight into space. The moon was always my favorite, but Saturn would knock me out. There it was, shimmering, the rings coming into focus showing rings within rings, bands on Saturn's disk, and subtle colors galore. I was no longer in Los Angeles, or even on Earth.

Griffith Observatory became the center of my universe when I was about 13. By that time, I had memorized the constellations, knew dozens of star names, and could point things out at night. In sixth grade, I joined the junior section of the Los Angeles Astronomical Society. Every now and then my dad took me to their monthly meetings where real astronomers like Allan Sandage and Olin Wilson lectured in the Zeiss planetarium. But every Friday, I boarded the Pico bus at 3:30 p.m., and, after three transfers I got to the Observatory in time for the telescope making shop to open at 5:30 p.m.

I was making a four-inch mirror for my first telescope. The kit cost under $10, which I could handle from my paper routes. It took a year of Fridays to make my mirror, and often the instructors would quiz us about astronomy. Sometimes we went to planetarium shows, and sometimes to the telescope after the grinding session. This was the fun of it.

The payoff was taking my telescope out in the backyard and starting to mind the heavens. "Star parties" became the high point of my existence in the summers. My dad had been taking me to them since I was about 11, but now I had something to offer. Again my dad would be enlisted and we would fill the station wagon with the telescope, a bit of camping gear and maps and food, and take off to one of the LAAS monthly meetings under the stars: informal picnics at night, usually 40 to 60 miles out of town, most

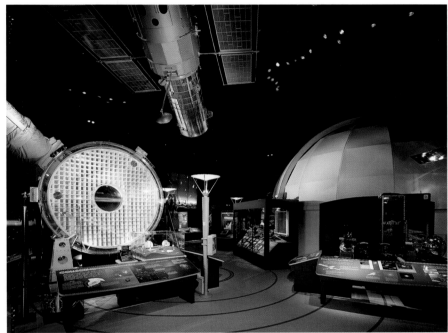

frequently in the mountain range behind Mount Wilson, totally out of sight of city lights. Several large parking areas would be filled with cars, telescopes, and people. The telescopes were magnificent. No formal lectures, but a lot of talk at every step about what was interesting to see, what could be seen with what size telescope, and tips on making better telescopes. I kept going to star parties when I could drive myself; with my friends these became all-night affairs, camping under the stars and talking astronomy.

Star parties abound today. Our museum participates almost year-round, advertising local events. Planetariums and museums are usually good places to start to find out where they might be taking place in your area.

I eventually became a guide and then school-show lecturer at Griffith, working my way through UCLA, and feel that these life experiences did more than anything to stimulate an indelible fascination with the universe as a personal, human enterprise. This, I feel, is the best way to explore the universe.

Explore the Universe at the National Air and Space Museum contains a wealth of astronomical instruments. Many objects in the exhibition are pictured in this book.

ABOUT THE AUTHORS

VON DEL CHAMBERLAIN is the director emeritus of Hansen Planetarium and is an expert on Native American skylore. He is the author of *When Stars Came Down to Earth: Cosmology of the Skidi Pawnee Indians of North America*.

DAVID DEVORKIN is the curator for history of astronomy and space science at the Smithsonian's National Air and Space Museum. He curated the *Explore the Universe* gallery. He has recently written *Henry Norris Russell: Dean of American Astronomers*.

MARGARET JOAN GELLER is a senior scientist at the Smithsonian Astrophysical Observatory. Her pioneering maps of the nearby universe uncovered patterns extending for hundreds of millions of light years. She plans deeper maps to explore the origin of these patterns. Background for her chapter can be found in *Mastery: Interviews with 30 Remarkable People*, by Joan Ames. *Before the Beginning: Our Universe and Others*, by Martin Rees, and *A Short History of the Universe*, by Joseph Silk.

OWEN GINGERICH is research professor of astronomy and the history of science at the Harvard-Smithsonian Center for Astrophysics. He has authored and edited numerous reviews, articles, and books, including *The Great Copernicus Chase and Other Adventures in Astronomical History*, *The Eye of Heaven: Ptolemy, Copernicus, Kepler*, and *An Annotated Census of Copernicus's de Revolutionibus*.

MICHAEL HOSKIN is a fellow of Churchill College, Cambridge, and formerly head of the department of history and philosophy of science, Cambridge University. He is the founder/editor of the *Journal for the History of Astronomy*. He has authored or edited numerous works, including *The Cambridge Illustrated History of Astronomy*.

CHRISTINE MULLEN KREAMER is a curator at the Smithsonian National Museum of African Art. She has written many works about Africa. For more information on African cosmology see *Icons: Ideals and Power in the Art of Africa* by Herbert C. Cole, *Tabwa: The Rising of a New Moon* by Evan F. Mauer and Allen F. Robers, and *The Anatomy of Architecture: Ontology and Metaphor in Batammaliba Architectural Expression* by Suzanne Preston Blier.

JOHN S. MAJOR is an independent scholar based in New York and concurrently Senior Lecturer at The China Institute. He formerly taught at Dartmouth College and is a past director of the China Council of the Asia Society. He is co-author with Joseph Needham, et al., of *The Hall of Heavenly Records*, and author of *The Land and People of China*, and of *Heaven and Earth in Early Han Thought*.

VERA RUBIN is an astronomer at the Carnegie Institution of Washington Department of Terrestrial Magnetism. She is a Presidential Medal of Science winner and is generally credited with proving the existence of "dark matter" in galaxies. She is the author of *Bright Galaxies Dark Matters*.

SARA SCHECHNER is the David P. Wheatland curator of historical scientific instruments at Harvard University. She has written *Comets, Popular Culture, and the Birth of Modern Cosmology*, and the introduction to *Western Astrolabes* by Roderick and Marjorie Webster.

ROBERT SMITH is a professor and chairman of the department of history and classics at the University of Alberta in Canada. He is the author of *The Expanding Universe: Astronomy's "Great Debate," 1900-1931* and *The Space Telescope: A Study of NASA, Science, Technology, and Politics*, which won the 1990 History of Science Society's Watson Davis Prize.

F. RICHARD STEPHENSON is a professor of astrophysics at the University of Durham in England. He is co-author (with David Clarke) of the pioneering work: *The Historical Supernovae*, which helped raise general awareness of the astrophysical value of historical data.

J. ANTHONY TYSON is a distinguished member of the technical staff at Lucent Technologies/Bell Labs with specialties in experimental gravitation and cosmology. He is presently researching techniques to develop observational probes of dark matter and dark energy in the universe. He is a leading proponent of the Dark Matter Telescope.

DEBORAH JEAN WARNER is the curator of the history of the physical sciences at the Smithsonian's National Museum of American History. She is founder and former editor of the journal *Rittenhouse* and has authored *The Sky Explored: Celestial Cartography, 1500-1800*.

DAVID WILKINSON is a professor of physics at Princeton University. He has been interested in measuring the cosmic background radiation for over a quarter of a century. He helped design the COBE satellite and recently has been a partner creating the MAP satellite, which was launched in June 2001 to collect observations for two years that will greatly refine our knowledge of how radiation and matter decoupled 300,000 years after the big bang.

ROBERT W. WILSON is an astronomer at the Harvard-Smithsonian Center for Astrophysics. He was the 1978 recipient of the Nobel Prize for Physics for his discovery, with Arno Penzias, of the cosmic microwave background radiation. His essay was adapted from earlier remarks, including a chapter in *Serendipitous Discoveries in Radio Astronomy*.

250

ILLUSTRATION CREDITS

Cover, National Air and Space Museum, Smithsonian Institution. Illustration by Hugh McKay/ mckayscheer.com; 1, National Air and Space Museum (NASM), Smithsonian Institution. Illustration by Matthew Frey/Wood, Ronsaville and Harlin, Inc.; 2-3, NASM, Smithsonian Institution. Illustration by Rob Wood/Wood, Ronsaville and Harlin, Inc.; 4-5, NASM, Smithsonian Institution. Illustration by Rob Wood/Wood, Ronsaville and Harlin, Inc.; 8-9, Rare Book Division, Library of Congress; 11, NASM, Smithsonian Institution. Illustration by Matthew Frey/Wood, Ronsaville and Harlan, Inc.; 13, Courtesy David DeVorkin; 14-15, THE GRANGER COLLECTION, New York; 18, Hulton|Archive/Getty Images; 19, Smithsonian Institution Libraries, The Dibner Library of the History of Science and Technology; 21, NASM, Smithsonian Institution. Photo by Eric Long; 23, Erich Lessing/Art Resource; 26, Erich Lessing/Art Resource; 27, Istanbul University Library; 28, Reunion des Musees Nationaux/Art Resource; 29, NASM, Smithsonian Institution. Photo by Eric Long; 31, Giraudon/Art Resource; 32, Reunion des Musees Nationaux/Art Resource; 33, THE GRANGER COLLECTION, New York; 37, A.M.S. Foundation for the Arts, Science and Humanities, courtesy of the Arthur M. Sackler Gallery, Smithsonian Institution, Washington, D.C.: MLS1802; 38, Courtesy of Adler Planetarium & Astronomy Museum, Chicago, Illinois; 39, THE GRANGER COLLECTION, New York; 41, Werner Forman Archive/Art Resource; 43, Jerry Schad/Photo Researchers; 45, Stephen Trimble; 46, David Brill; 46-47, George H.H. Huey; 48, Stephen Trimble; 49, 'The Raven and the First Men" from the collections of the University of British Columbia Museum of Anthropology. Photo by Gunther Marx Photography/COR-BIS; 51, Courtesy Von Del Chamberlain. Reproduced with permission of the Pawnee Nation of Oklahoma; 52, Drew Possessky/stosh@ptd.net; 53, Courtesy Von Del Chamberlain/Artwork by Clifford Jim; 55, Tony & Daphne Hallas/Science

Photo Library/Photo Researchers; 56-57, Kenneth Garrett; 59, Kenneth Garrett; 61, Fred Espenak (www.MrEclipse.com); 63, Jose Azel/Aurora; 65, National Museum of African Art, Smithsonian Institution. Photo by Franko Khoury, ID#84-6-6.1; 67, NASA/CXC/SAO; 69, National Museum of African Art, Smithsonian Institution. Photo by Franko Khoury, ID#83-3-8; 71, National Museum of African Art, Smithsonian Institution. Photo by Franko Khoury, ID#91-22-1; 73, THE GRANGER COLLECTION, New York; 75, Art Resource; 77, NASM, Smithsonian Institution. Photo by Eric Long; 78, Bibliotheque Nationale de France; 81, Smithsonian Institution Libraries, The Dibner Library of the History of Science and Technology; 82-83, Courtesy of Adler Planetarium & Astronomy Museum, Chicago, Illinois; 85, Erich Lessing/Art Resource; 87, Courtesy Owen Gingerich; 88, THE GRANGER COLLECTION, New York; 91, Library of Congress, Geography and Map Division; 93, NASM, Smithsonian Institution. Photo by Eric Long; 94, Archiv/Photo Researchers; 96-97, Hansen Planetarium, Salt Lake City; 98-99, NASM, Smithsonian Institution.; 101, NASM, Smithsonian Institution. Illustration by Rob Wood/Wood, Ronsaville and Harlin, Inc.; 103, P.K. Chen/Yerkes Observatory; 104, Art Resource; 105, THE GRANGER COLLECTION, New York; 106 (both), By permission of the Syndics of Cambridge University Library/ Newton Manuscripts, MS. 3965, folio 280r and folio 74r; 109, Smithsonian Institution Libraries, The Dibner Library of the History of Science and Technology; 113, NASM. Photo by Eric Long; 114 (le), By courtesy of the National Portrait Gallery, London; 114 (rt), THE GRANGER COLLECTION, New York; 115, Library of Congress; 116, Smithsonian Institution Libraries, The Dibner Library of the History of Science and Technology; 117 (le), Courtesy David DeVorkin; 117 (rt), Birr Scientific Heritage Foundation; 118-119, Science Museum/Science and Society Picture Library; 121, The Observatories of the Carnegie Institution of

Washington; 122-123, The Observatories of the Carnegie Institution of Washington; 124, NASM, Smithsonian Institution. Photo by Eric Long; 125, The Observatories of the Carnegie Institution of Washington; 127, David Parker/Science Photo Library/Photo Researchers; 128, Lowell Observatory; 131, Courtesy David DeVorkin; 132, NASA; 133, NASM, Smithsonian Institution. Photo by Eric Long; 134-135, R. Williams and the HDF Team (STScI), NASA; 137, NASM, Smithsonian Institution. Artwork by Keith Soares/Bean Creative; 139, NASA and The Hubble Heritage Team (STScI/AURA); 140, Courtesy David DeVorkin; 141, NASM, Smithsonian Institution. Artwork by Keith Soares/Bean Creative; 143, NASM, Smithsonian Institution. Photo by Eric Long; 145, NASM, Smithsonian Institution. Photo by Eric Long; 147, Courtesy David DeVorkin; 148, Science Photo Library-Dr. F.D. Miller/Photo Researchers; 149, AURA, NOAO, NSF; 150, AIP Emilio Segre Visual Archives, Physics Today Collection; 151, This item #COPC9 is reproduced by permission of The Huntington Library, San Marino, CA; 152-153, NASM, Smithsonian Institution; 155, NASM, Smithsonian Institution. Illustration by Rob Wood/Wood, Ronsaville and Harlin, Inc.; 157, NASM, Smithsonian Institution. Illustration by Rob Wood/Wood, Ronsaville and Harlin, Inc.; 159, Courtesy F. Richard Stephenson; 160, J. Hester and P. Scowen (ASU), NASA; 161, S. L. Snowden (NASA/GSFC/LHEA); 162, Courtesy F. Richard Stephenson; 165, Dr. Christopher Burrows, ESA/STScI and NASA; 166, Courtesy of Adler Planetarium & Astronomy Museum, Chicago, Illinois; 167, Courtesy F. Richard Stephenson; 169, Dr. Eric Gotthelf (Columbia); 170-171, Space Telescope Science Institute/NASA/Science Photo Library; 173, Graphics by Emilio Falco, Smithsonian Astrophysical Observatory and Mark Bajuk, National Center for Supercomputing Applications; 177, Graphics provided by Michael J. Kurtz, Smithsonian Astrophysical Observatory; 179, (top left) Courtesy

251

David DeVorkin; (center left) NASM, Smithsonian Institution. Illustration by Rob Wood/Wood, Ronsaville and Harlin, Inc.; (bottom left) NASM, Smithsonian Institution. Artwork by Keith Soares/Bean Creative; (top right) NASM, Smithsonian Institution. Illustration by Matthew Frey/Wood, Ronsaville and Harlin,. Inc.; (center right) NASM, Smithsonian Institution. Illustration by Rob Wood/Wood, Ronsaville and Harlin, Inc.; (bottom right) Graphics provided by Michael J. Kurtz, Smithsonian Astrophysical Observatory; 180, Virgo Consortium; 181, Virgo Consortium; 182, VLT UTI + FORS1; 184-185, Photograph provided by Daniel G. Fabricant, Smithsonian Astrophysical Observatory; 187, Roger Smith/NOAO/AURA/NSF; 189, Adam Block/AURA/NOAO/NSF; 193, Simulation performed by Darren Reed with PKDGRAY, written by Joachim Stadel & Tom Quinn, on SGI Origin 2000's at NCSA & NASA Ames; 194, Courtesy Vera Rubin; 196, Courtesy Peter Wannier/Graph based on article originally appearing in the PUBLI-CATIONS OF THE ASTRONOMI-CAL SOCIETY OF THE PACIFIC (Kent, 1989, PASP, 101, 489) © 2000, Astronomical Society of the Pacific; reproduced with permission of the Editors; 197, Malin/CALTECH 2001; 199, The Boomerang Collaboration; 203, Peter Barthel (Kapteyn Inst.) et al., FORS1, VLT ANTU, ESO; 205, Greg Kochanski (Lucent Technologies, Inc./Bell Labs); 208-209, NASM, Smithsonian Institution. Artwork by Keith Soares/Bean Creative; 210, Tony Tyson (Lucent Technologies, Inc./Bell Labs); 211, NASM, Smithsonian Institution. Photo by Eric Long; 212, Ed Turner & Wes Colley (Princeton U.), and Tony Tyson (Lucent Technologies, Inc./Bell Labs) and NASA; 213, Greg Kochanski & Tony Tyson (Lucent Technologies, Inc./Bell Labs) and Ian Dell'Antonio (Brown U.); 214, Ian Dell'Antonio (Brown U.) and Tony Tyson (Lucent Technologies, Inc./Bell Labs) and NASA; 215, Ayana Holloway & Tony Tyson (Lucent Technologies, Inc./Bell Labs); 216, David Wittman & Tony Tyson (Lucent Technologies, Inc./Bell Labs) and NSF; 217, Tony Tyson (Lucent Technologies, Inc./Bell Labs); 219, Roger Angel & Warren Davison (U. of Arizona,

Steward Observatory), and The Astrophysics and Precision Engineering Groups (Lawrence Livermore National Labs); 221, NASM, Smithsonian Institution. Illustration by Hugh McKay/mckayscheer.com; 224-225, NASM, Smithsonian Institution. Artwork by Keith Soares/Bean Creative; 227, Lucent Technologies, Inc./Bell Labs; 228, Lucent Technologies, Inc./Bell Labs; 230-231, Lucent Technologies, Inc./Bell Labs; 233, Mullard Radio Astronomy Laboratory/Science Photo Library/Photo Researchers; 235, Courtesy Dr. David Wilkinson, Princeton University, Dept. of Physics, Joseph Henry Labs; 236, Courtesy Dr. David Wilkinson, Princeton University, Dept. of Physics, Joseph Henry Labs; 239, NASM, Smithsonian Institution. Photo by Eric Long; 240, NASM, Smithsonian Institution. Photo by Eric Long; 241, Courtesy Dr. David Wilkinson, Princeton University, Dept. of Physics, Joseph Henry Labs; 243, Mehau Kulyk/Science Photo Library/Photo Researchers; 244, COBE, MAP, NASA; 245, COBE, MAP, NASA; 246-247, The Boomerang Collaboration; 249, NASM, Smithsonian Institution. Photo by Eric Long.

Credits for quotations in Chapter 4, "Africa's Moral Universe:" 60, Anthony Aveni, *Conversing with the Planets: How Science and Myth Invented the Cosmos.* New York: Times Books, 1992, pp. xi; 64, Yoruba praise poem for Shango translated by Robert Thompson in *Flash of the Spirit.* New York: Vintage Books, 1984; 64, Traditional Bamana proverb (Mali); 68, Traditional Jukun proverb (Nigeria); 69, Yoruba praise poem recorded by C.L. Adeoye, translated by Babatunde Lawal, "À yà gbó, à yà tó: New Perspectives on Edan Ògbóni" in *African Arts* 28 (1): pp. 36-49, 98-100.

254

BEYOND EARTH
Mapping the Universe

Published by the National Geographic Society

John M. Fahey, Jr., *President and Chief Executive Officer*

Gilbert M. Grosvenor, *Chairman of the Board*

Nina D. Hoffman, *Executive Vice President*

Prepared by the Book Division

Kevin Mulroy, *Vice President and Editor-in-Chief*

Charles Kogod, *Illustrations Director*

Marianne R. Koszorus, *Design Director*

Staff for this Book

David DeVorkin, *Editor*

Marlis McCollum, *Text Editor*

Melissa G. Ryan, *Illustrations Editor*

Lyle Rosbotham, *Art Director*

Lisa Krause Thomas, *Assistant Editor*

Joan Mathys, *Photo Researcher*

David Romanowski, *Picture Legends Writer*

R. Gary Colbert, *Production Director*

Richard S. Wain, *Production Project Manager*

Sharon Kocis Berry, *Illustrations Assistant*

Deborah E. Patton, *Indexer*

Manufacturing and Quality Control

George V. White, *Director*

Clifton M. Brown, *Manager*

Phillip L. Schlosser, *Financial Analyst*

One of the world's largest nonprofit scientific and educational organizations, the National Geographic Society was founded in 1888 "for the increase and diffusion of geographic knowledge." Fulfilling this mission, the Society educates and inspires millions every day through its magazines, books, television programs, videos, maps and atlases, research grants, the National Geographic Bee, teacher workshops, and innovative classroom materials. The Society is supported through membership dues, charitable gifts, and income from the sale of its educational products. This support is vital to National Geographic's mission to increase global understanding and promote conservation of our planet through exploration, research, and education.

For more information, please call 1-800-NGS LINE (647-5463) or write to the following address:

National Geographic Society
1145 17th Street N.W.
Washington, D.C. 20036-4688 U.S.A.

Visit the Society's Web site at www.nationalgeographic.com.

Library of Congress Cataloging-in-Publication Data

Beyond Earth: Mapping the Universe
                    p. cm.
Includes biographical references and index.
ISBN 0-7922-6467-3
1. Cosmology—History. I. National Geographic Society (U.S.)

QB981 .B56 2002
523.1'09—dc21